SPACE
2069

SPACE 2069

2069

AFTER APOLLO:
BACK TO THE MOON,
TO MARS,
AND BEYOND

DAVID WHITEHOUSE

ICON

First published in the UK and USA in 2020
by Icon Books Ltd, Omnibus Business Centre,
39–41 North Road, London N7 9DP
email: info@iconbooks.com
www.iconbooks.com

This edition published in the UK and USA in 2021 by Icon Books Ltd

Sold in the UK, Europe and Asia
by Faber & Faber Ltd, Bloomsbury House,
74–77 Great Russell Street,
London WC1B 3DA or their agents

Distributed in the UK, Europe and Asia
by Grantham Book Services, Trent Road, Grantham NG31 7XQ

Distributed in the USA
by Publishers Group West,
1700 Fourth Street, Berkeley, CA 94710

Distributed in Canada by Publishers Group Canada,
76 Stafford Street, Unit 300
Toronto, Ontario M6J 2S1

Distributed in Australia and New Zealand
by Allen & Unwin Pty Ltd, PO Box 8500,
83 Alexander Street, Crows Nest, NSW 2065

Distributed in South Africa
by Jonathan Ball, Office B4, The District,
41 Sir Lowry Road, Woodstock 7925

Distributed in India by Penguin Books India,
7th Floor, Infinity Tower – C, DLF Cyber City,
Gurgaon 122002, Haryana

ISBN: 978-178578-719-5

Typeset in Walbaum MT by Marie Doherty

Printed and bound in Great Britain by
Clays Ltd, Elcograf S.p.A.

For Jill

'By three methods we may learn wisdom: first, by reflection, which is the noblest; second, by imitation, which is the easiest; and the third by experience, which is the bitterest.'

—CONFUCIUS

CONTENTS

ACKNOWLEDGEMENTS

I recall one Tuesday afternoon in the 1990s I was sitting at my desk in Broadcasting House, home of the BBC, and the phone rang.

'Hello, it's Arthur Clarke here. Have you seen the latest images from Mars? Most weird.'

Arthur C. Clarke wanted there to be life on Mars. He felt it would speed up getting humans there.

'Dammit,' he told me, 'I thought we'd be at Jupiter by 2001, but I don't think we will be back on the Moon by then.'

Those who know his works would say that Arthur was far too optimistic over timescales of decades and centuries, but far too pessimistic when looking thousands or millions of years into the future! It was one of Arthur's books that really inspired me when I was barely in my teens. *The Promise of Space*, it was called. I still have it. I wonder what Arthur would have said if I had the chance to tell him that one day I would write a book about 'only' the next 50 years in space.

This book grew out of my previous book, *Apollo 11: The Inside Story*. It's a kind of sequel. Having just looked back 50 years to the first lunar landing I thought it would be interesting to assess what we might have achieved 100 years after the 'small step'. While writing it I realised what Arthur would have felt like when writing his profiles of the future. We will never know what we got right.

I recall talking with Carl Sagan in a hotel in London about the future of space and our discussion automatically moved to Mars. He also talked, in a typically Sagan way, of looking for a benign aperture through which to see the 21st century and a hopeful future for the human species. He was aware that we could take our Earthly troubles into space.

Then there was Patrick Moore, who wrote a fascinating book in the 1970s called *The Next 50 Years in Space*, with wonderful illustrations by David Hardy. It had spaceships landing on the moons of Jupiter and astronauts driving pressurised rovers on the moons of even more distant Saturn. The book is ever wonderful but I did feel regret on realising humanity would probably never do those things. Patrick's response was to say, 'Why don't we get the telescope out and look for sunspots?'

I feel this book might disappoint Arthur, Carl and Patrick. Arthur would perhaps have me include a mysterious alien artefact travelling through the solar system as in his 1978 novel *The Fountains of Paradise*. Carl would be disappointed I didn't tackle his favourite subject, the search for intelligent life in space, which I have left out to concentrate on the human story, knowing it's a discovery that could be made tomorrow or never. It's a subject for another book. Patrick, in a parallel universe, would have got his wish to present his TV show *The Sky at Night* from the surface of the Moon. This book explains why he sadly never got the chance.

ACKNOWLEDGEMENTS

So many people over so many years have contributed to this book through correspondence and conversations. I would like to thank Nick Booth for valuable comments and Icon's Robert Sharman for his expert editing that has significantly enhanced this book. My agent Laura Susijn has been a wonder, even though at times I must have driven her to distraction. My wife Jill has been as marvellous as ever, and my children Christopher, Lucy and Emily have been so enthusiastic. Lastly, thanks to Tobi our Cavalier King Charles Spaniel who has on long walks through the countryside helped me sort out many problems, and frequently brought me down to Earth.

David Whitehouse
Hampshire
June 2020

PREFACE

In 2012 the Austrian skydiver Felix Baumgartner jumped from the stratosphere, launching himself from a capsule suspended from a helium balloon. Before he jumped, he looked at the Earth below him, its curvature clearly visible, the raw sunlight showing him and his capsule in a brutal light. His pressure suit was custom-made and had four layers to keep him protected from the harsh environment around him. He looked every inch a spaceman.

Baumgartner was at an altitude of 38,969 metres in the very upper reaches of the atmosphere, which was cold, thin and deadly. Yet as he contemplated the moment, the atmosphere around him was warmer and thicker than that found on the surface of Mars.

PART 1

BACK TO THE MOON

EXILES

———•———

The Moon.
Shackleton crater.
Lunar latitude 90 degrees south.
Date: 2069.

At the exact moment of the centenary, the depths of Shackleton crater at the Moon's south pole went dark, as they had been for billions of years before humans arrived. The eight giant mirrors on the crater's rim – heliostats that rotated to keep the Sun's light reflected into the crater – started to turn and the pools of light they cast moved swiftly across the crater's floor, up its ramparts, disappearing into space. The mirrors turned to the precise orientation required to reflect the Sun's light towards Earth: to a particular spot in North America. Millions on Earth were watching the southern region of the Moon's grey disc, whether with the unaided eye, binoculars or telescope, hoping to catch the optical flare from the mirrors on the Moon.

3

Wapakoneta had a population of 12,236 at the 2060 census but the town had swelled for the event. Teachers and students from the high school were waving their red-and-white school flags as well as the Stars and Stripes. All over the globe if people were not looking at the Moon, or couldn't see it, they were watching the celebrations. They all cheered as they saw the flashes from the Moon, saluting the birthplace of the first man to walk on it, Neil Armstrong, 100 years ago. At the time of the 50th anniversary there were only four of the original moonwalkers alive and only 20 per cent of the population had been around at the time of the first Moon landing. Fifty years later no one who took part in the Apollo program was alive. Given the advances in medicine there were still many who recalled the event. But there was a letting it slip into history – the loss of the Apollo heroes and anyone who had ever known them personally. But there were moonwalkers at the centenary of Apollo 11. The thirteenth person on the Moon was in her eighties and just as sprightly as Buzz Aldrin had been in 2019.

A few minutes later attention turned to the screens and a large black-and-white image of the lunar surface. This was the view from the cameras positioned around the First Footprint sanctuary. The main cluster of cameras was stationed on the edge of West crater about 400 metres away. Nothing was allowed closer, but the view from the tower was clear. The desolate lower stage of the Apollo 11 lunar module and the flag lying on the ground. Floodlights highlighted with crisp shadows every footprint and scuff mark. The tracks leading towards Little West crater, 50 metres east of the lunar module, were visible. It was an unplanned excursion, when Armstrong had gone to get a look inside near the end of the two-and-a-half hours that he and Aldrin had spent on the surface. The image closed in on the lunar module leg where the very first footfall was made and mostly obliterated by subsequent boot marks. The site had been laser scanned and converted to virtual reality so anyone could walk up to the forlorn module and read the plaque: 'Here men from the planet Earth …'

But not everyone celebrated the same way, and not everyone on the Moon was thinking back to Apollo 11. There were three sets of broadcasts from the Moon on that day of celebrations. Shackleton base scientists had answered questions from the world's media and from schools around the world. The secondary lunar outpost positioned near the Moon's equator took viewers on a tour of their strange underground habitat. The far larger Chinese base, in the northern polar region and as far away from Shackleton as it was possible to get on the Moon, made its contribution, but it was mainly for their own country, given the international tensions. There was no broadcast from their other base. It was also watched by the military space chiefs of both countries as they kept a watchful eye on the space around the Moon, for there had been times of conflict on and around the Moon.

The heliostats moved once more, this time to reflect the sunlight across the inner solar system towards Mars on a 28-minute journey to the red planet, and the Mars Optical Telescope – the only one on the red planet – turned its gaze towards the Moon from the dusty and desolate floor of the

Valles Marineris. The Chinese base at Acidalia Planitia ignored the signal.

The celebrations of the 100th anniversary of the landing of Apollo 11 on the Moon were subdued at the International Mars base despite the optimism on display in the Earth–Moon–Mars link-up and the messages from the world's leaders. Events happening 500 million kilometres away inevitably seemed remote from the viewpoint of honorary Martians. For some it reminded them, as if they needed reminding, that they depended on Earth for their survival, being always just two resupply trips away from extinction. The Martians, eighteen of them in the International Mars Colony, eight at the Chinese base – many fewer than at the lunar colonies – called themselves Martians, though they had all been born on Earth and it was to Earth they would eventually return, even if they could never really feel at home there after their time on an alien planet. Privately some of them knew in their heart of hearts that they could not face the voyage home, and that eventually they would join the other sealed graves on Mars.

A hundred years after Apollo 11, mankind had journeyed into the solar system and faced a new barrier, one that would probably take another hundred years to overcome, if overcome it could be. From Mars we stared out towards the asteroid belt and beyond to the gas giants of Jupiter and Saturn, and the ice giants of Uranus and Neptune in the cold, dark outer reaches of the solar system beyond. Then in the century after the Apollo centenary we could imagine a voyage into the asteroid belt, perhaps overseeing the artificial-intelligence swarms that roamed among these rocky bodies. But we could go no further. The vast distances and the long durations of the flights were too much at present, let alone the radiation. The humans that would go out there would be different. Modified, enhanced, resilient and protected in a way that space travellers had thus far not been. A human voyage to Jupiter and its remarkable moons, to Saturn's moon Titan and the beguiling Enceladus belong to the centuries to come, and to different people. Looking inward towards the Sun, we cannot live on Venus or Mercury. For humans Earth marks the inner boundary of our reach into the solar system.

The next 50 years will see the start of our divergence. By the end of it the Moon and Mars will have their own people for whom Earth has never been their home. Some will become exiles, unable to visit Earth because its gravity would kill them. Some will become a new branch of humanity, regarding themselves liberated from the confines of the planet of their predecessors.

Encompassing the Moon and Mars will not just be about the journey, the technology of travel and survival in space. It will include all of the science we will discover in the next five decades. Better control of our bodies and brains, our new attitudes, our new and ancient fears – which perhaps are the same. Space colonists will not be the Mayflower pilgrims of the 21st century, looking for release from old ways and oppression. We will take our tyrannies with us, along with our tragedies, fears and hopes. For the next 50 years we will take our Earth thinking with us, reflecting and amplifying the politics of our home planet, perhaps acting out its battles.

We humans begin our expansion outwards, in the first phase to the Moon and Mars. This is what concerns us here.

✳

It is now August 2024 and two astronauts are flying over the lunar surface for the first time in many decades, travelling a course long abandoned. Passing below them is the Moon's most prominent crater, mostly covered in shadow as the morning Sun, striking its western rim, moves down its flanks, unveiling its jumbled floor of cracks, small hills and domes. Tycho is at a lunar latitude of 43° S and is 85 kilometres wide. Its signature streaks which span the entire Moon and which are so prominent at certain Sun angles are not easily seen by the crew. The last of the Surveyor soft landers is resting just 40 kilometres from the rim, having been there since 1968 when it was landed to test the stability of the surface. It didn't disappear into dust. Sitting on what was technically called the impact melt ejecta blanket, its cameras could see Tycho on the horizon. Apollo 20 had been due to land next to it sometime in the mid-1970s, but had been cancelled before any detailed plans could be made.

Now, the radar registers the crater's cliff faces that are higher than the Grand Canyon, as well as the terraced and slumped terrain that guards Tycho's dramatic heart. Neither of the astronauts looks towards its dramatic central peak, six times the height of the Empire State Building, and the enormous boulder sitting on it – one of nature's tricks played on the Moon – but they do reference it in passing.

'Site of TMA-1,' says one. Both of them know she is referring to the black monolith left behind by enigmatic aliens in the film *2001: A Space Odyssey*.

'We are a little late getting here,' says the other.

The pair are part of a team nicknamed the 'Turtles' – NASA's 22nd astronaut selection: twelve Americans (one subsequently resigned) and two Canadians. They were chosen from over 18,000 applicants in June 2017 and started a two-year training course. In January 2020 they were assigned to NASA's Artemis program. Most, perhaps all of them, will walk on the Moon.

To get there the two moonwalkers will, with two others, be launched into space using the super rocket of the US Space Launch System and the Orion capsule, which looks like the Apollo Command Module but is larger, more complicated and more capable. At one time it was planned to have them boosted out of Earth orbit for a five-day trip to the Lunar Gateway – a simple space station in a very elliptical orbit around the Moon. From there the chosen two would have entered the Lunar Descent Vehicle, undocked, fired its thrusters and begun a twelve-hour descent to the surface. In March 2020 NASA changed that plan. Although the Gateway project was still to go ahead, it was decided that the landing mission would not dock with it but would head directly to the lunar surface. The change was made to save money and time, given the uncertainties introduced into the project because of the coronavirus pandemic. But the astronauts will not be going back to an Apollo landing site, or anything like it.

In the darkness of their capsule their eyes are on their dimly illuminated instruments: altitude, rate of descent, velocity, range, fuel. The infra-red laser radar picks up its reflection from the beacon on the lunar rover already placed at the landing site.

When he came in for the first landing on the Moon 55 years previously, Neil Armstrong had nothing like the information presented to this crew. They have screens showing all the spacecraft's vital signs, a detailed map and profile of the Moon along with their trajectory, as well as excellent comms. Back then Armstrong's hands had been curved around two joysticks as he leant forward, peering below through the overhanging window of the lunar module. He was flying for his life, calling from memory details of the terrain before him, looking to avoid boulders, hearing mission control through the static, monitoring the fuel supply: 'Low level,' they said. He had Buzz Aldrin next to him calling out their altitude and descent: 'Four forward, four forward, drifting to the right a little.'

Armstrong was landing on a very different part of the Moon – the Plains of Tranquillity. The crew of Artemis 3 are flying over the Moon's most rugged terrain. The ground underneath them is becoming darker, more shadowed as they head poleward, exactly the opposite of what Armstrong encountered. Artemis 3 is travelling yet deeper into the shadows beneath. No one has ever been this way before.

DRIFTING

———•———

Someone once said – exactly who is not known for sure; it's been attributed to movie mogul Sam Goldwyn and the physicist Niels Bohr among others – that predictions are difficult, especially about the future! It's true about space activity. If you were celebrating the first manned landing on the Moon in 1969 you would have not predicted where we are 50 years later, with no one having been to the Moon since 1972. Back then for some it was a time of optimism. Spiro Agnew, vice-president to President Nixon, said men would be on Mars by 1984. Instead we got the first untethered space-walk, space shuttle Discovery's maiden voyage and the release of *The Terminator*. The famous movie co-written by Stanley Kubrick and Arthur C. Clarke came out in 1968. It depicts a large spacecraft heading to Jupiter. You know when it was set – 2001. We haven't come anywhere near where our space dreams once imagined we would be. The changes that have happened to our society, our technology and ourselves were poorly predicted. Changes have been faster in some areas, slower in many others.

In another 50 years, many things will have changed: our environment will be different; our bodies will certainly change as the result of new medical technology; our reach may expand or contract; our optimism ... well, we can hope.

When I was a schoolboy, around the time of the first Moon landing, I imagined that within a few years we would have space stations and fabulous vehicles that would make travelling to them routine, allowing many of us to become astronauts. Fifty years after Apollo we might have hotels in space, settlements on the Moon and colonies on Mars. Perhaps we would travel even further, to Jupiter and beyond, just like the Discovery One with Dr David Bowman and Dr Frank Poole, and their conscious artificial intelligence HAL. This was the future I anticipated and, as youthful optimism faded, watched slip away, my space dreams receding a little further each year.

Fifty years after Apollo fewer than 600 people had ventured into space and only 24 of them beyond Earth orbit. But there are signs that the stagnation is ending. We will be back on the Moon, tourists will dip their toes into space and the infrastructure of space will continue to grow and touch almost everyone on Earth with communications, navigation, transport.

Ever since the great speeches of President John F. Kennedy setting the US on course for the Moon before the 1960s were out, NASA – the US Space Agency – has wanted another JFK moment, another impetus to move outwards that the politicians who control the purse strings could get behind. Indeed, every subsequent president has wanted his own Kennedy moment, a speech as memorable and inspiring as JFK's.

Since the heady days of Apollo there have been three times when a US president has directed the nation to go back to the Moon, but none of them ever got anywhere. Presidential initiatives and administrations came and went. At times the goal was the Moon, then it was Mars, then back to the Moon, then an asteroid, and then back to the Moon again. It almost seemed as if we didn't really want to go: we would go through the motions, make designs and spend billions building rockets and other things, but it never really felt that the end of the process would be footprints on the Moon. We resigned ourselves that each initiative would end in some sort of failure, consoling ourselves that starts and stops were part of the process, that future generations would benefit and that eventually the politicians would get it. Looking at why this happened is instructive and provides some idea of the problems those with moondust or Mars-dust in their eyes have when dealing with the political world. Neil Armstrong knew that you didn't get to the Moon with rockets and willpower. Any Moon mission is launched from the real and messy world.

After the Apollo 11 landing a Space Task Group was created by President Nixon to look into what to do next. One thing was clear to Nixon: it had to be much cheaper than the Apollo program, in fact no more Apollos. It consisted of Vice-President Spiro Agnew, NASA Administrator Thomas Paine, and the Secretary of the Air Force Robert Seamans. They came up with a space station, a base on the Moon, further exploration of Mars and a space shuttle. It was a bold plan, but it went against the political climate of the time.

These were the years of the poor people's marches and the Tet offensive. Paine was worried by the view of the Nixon transition team on space, which wanted to limit NASA's ambitions. They wanted NASA to continue exploring the Moon, but not Mars. Forget Mars. Paine wanted Vice-President Agnew to endorse his wider goals and thought that he could persuade Nixon. Paine was wrong: unlike his predecessor Lyndon B. Johnson, Nixon was not keen on space and to him Agnew's views were irrelevant.

In retrospect, Nixon put his own political interests ahead of a Kennedyesque space goal. What could he direct NASA to do that was anywhere as good as getting to the Moon? A trip to Mars would only benefit future presidents. He would give speeches about space exploration, invite the returning lunar astronauts to receptions at the White House and toast them, but space was not important to him politically. He gave a speech that held out a promise of the future, and then took it away.

> I have decided today that the United States should proceed at once with the development of systems and technologies designed to take American astronauts on landing missions to Mars. This system will center on a new generation of rockets, exploiting nuclear power, which will revolutionize and render routine long-haul interplanetary flights.
>
> The year 1971 was a year of conclusion for America's current series of manned flights to the Moon. Much was achieved in the three successful

landing missions – in fact, the scientific results of the third mission have been shown to greatly outweigh the return from all earlier manned spaceflights, to Earth orbit or the Moon. But it also brought us to an important decision point – a point of assessing what our space horizons are as Apollo ends, and of determining where we go from here.

[...]

We will go to Mars because it is the one place other than our Earth where we expect human life to be sustainable, and where our colonies could flourish. We will go to Mars because an examination of its geology and history will reflect back on a greatly deepened understanding of our own precious Earth.

Above all, we will go to Mars because it will inspire us to clearly look beyond the difficulties and divisions of today, to a better future tomorrow.

'We must sail sometimes with the wind and sometimes against it,' said Oliver Wendell Holmes, 'but we must sail, and not drift, nor lie at anchor.' So with man's epic voyage into space – a voyage the United States of America has led and still shall lead. Apollo has returned to harbor. Now it is time to swiftly build new ships, and to purposefully sail farther than our ancestors could ever have dreamed possible.

Throughout the triumph of Apollo there was never a time when the majority of Americans thought it was all worthwhile. Nixon knew that. He dismissed the Moon and Mars initiatives, and only endorsed the space station and the shuttle because he thought it would win him votes in California. Less than a year after the first footprint on the Moon, Nixon was even on the verge of cancelling Apollos 15, 16 and 17, which would have resulted in only six people walking on the Moon – half the eventual number. Despite Agnew's public statements about sending humans to Mars by 1984, dreams of returning to the Moon and going on to Mars faded while NASA struggled to get the space shuttle flying on an inadequate budget, making compromises in its design that would come back to haunt them in the future. Even during the tenth anniversary celebrations of Apollo 11, the White House incumbent President Carter showed he wasn't interested in the Moon and promoted his energy policy instead. He said it was 'neither feasible nor necessary' to commit to an Apollo-style space program, and his space policy included only limited, short-range goals.

In 1981 when the first space shuttle was launched, going back to the Moon, let alone to Mars, seemed a long way away. With each infrequent mission the gap between what the shuttle could do and what had been promised seemed to widen. I recall in the early years of the 1980s it was the shuttle or nothing. Moon return studies were discouraged, and Mars was almost a dirty word at NASA – or at least one seldom said, and then usually in hushed tones. But there was an undercurrent. Some scientists started to call themselves 'Mars Underground'. Studies carried out on the sidelines fuelled the excitement. Off-budget meetings were held, questions formulated, and missions imagined. Some were vocal about going to Mars, especially Carl Sagan. Despite this, official talk of returning humans to the Moon and sending them to Mars disappeared for almost a decade as NASA concentrated on getting the shuttle operating.

In 1984 President Ronald Reagan, who mostly saw space in military, intelligence and national security terms – hence his Strategic Defense Initiative (Star Wars) in 1983 – decided that the US needed to cooperate with its allies in space. That way costs would be saved, and risks shared, but years passed with very little happening. Eventually the US House of Congress got fed up with this and created a National Commission on Space to look at the impasse. Its chair was to be highly respected former NASA Deputy Administrator George Low, but he died before he could take up the post. Thomas Paine got the job, bringing with him his frustrations about Nixon's Space Task Group. Paine used to say that the nation spent more on Space Invaders than the space shuttle, and more on pizza than the entire space program. But Paine was not a wily Washington political operator, as was obvious from his Nixon days, and the eventual report, 'Pioneering the Space Frontier', was predictable and not detailed enough. Sure, it talked about lunar bases, a large space station, missions to Mars and nuclear-powered spacecraft, but to the politicians it read more like science fiction than a reasonable way forward.

The House of Congress knew the National Commission on Space report was not politically or financially acceptable and asked the new NASA Administrator, James Fletcher, for a more realistic vision. Fletcher understood the political realities better than Paine and gave the task to former astronaut Sally Ride, who had recently worked on the investigation into the space shuttle Challenger accident. She produced another report, 'Leadership and America's Future in Space', which this time addressed the differences between dreams and reality. She identified four possible directions for the United States: a human mission to Mars, a lunar base, robotic planetary exploration, and a 'Mission to Planet Earth', to study our home world. The one thing that caught on in Washington was the Mission to Planet Earth idea. Moon and Mars advocates were once again frustrated.

Ride worked in the Office of Exploration, which had very few staff. Soon Ride retired to private life and Aaron Cohen, from the Johnson Space Center (JSC) in Houston, took over. JSC had many studies for lunar bases and all sorts of future projects and so, for the next two years the office, known as 'Code Z', carried out its own studies of Moon and Mars missions. They included a large new booster based on the shuttle, known as 'Shuttle Z'.

In February 1988, the Reagan administration called for the 'expansion of human presence beyond Earth orbit', but Reagan was leaving office that year and things stalled again under President George H.W. Bush. Eventually the new administration tried to get things moving with the creation of another new body, a National Space Council, which was overseen by Vice-President Dan Quayle. Quayle had a bad image and was often a figure of ridicule. His advisors suggested that backing a large-scale space initiative would be good for his image. The Space Council idea had been suggested before, but Reagan hated it, even vetoing it on one occasion. His space outlook was influenced by the Department of Defense and he didn't want any Congressional interference. But President Bush was different and he liked the plan partly because it would give Quayle a public role in policy-making and help to dispel negative comments about his vice-president.

On the 20th anniversary of the first human landing on the Moon, President Bush, standing on the steps of the National Air and Space Museum with the crew of Apollo 11 by his side, proposed a long-range exploration plan encompassing a space station, a permanent return to the Moon, and a human mission to Mars. It became known as the Space Exploration Initiative (SEI). President Bush said the newly recharged National Space Council would provide ways to meet these objectives. He later set a 30-year goal for a crewed landing on Mars. If it had been met, humans would have been walking on Mars by 2019, the recently passed 50th anniversary of the Apollo 11 lunar landing.

But within a few short years even the SEI faded into history amid a bitter political war fought on many fronts. The failure of SEI, as well as the farce of the Hubble Space Telescope's flawed mirror and the problems in operating the space shuttle, its delays and its staggering cost, tarnished NASA's image. There were clashes both political and personal between the Space Council's Dan Quayle and Mark Albrecht versus NASA and its Administrator, the former astronaut Richard Truly; also between the White House, Congress and the Johnson Space Center's director Aaron

Cohen against, well, almost everyone. It all did the space effort little good.

The Space Council's real driving force was its secretary, Mark Albrecht who began looking for a new project to invigorate the civilian space program. Albrecht and the Space Council produced three proposals which looked just like the ones made before: a return to the Moon, a human mission to Mars, or both. But tension was building between the National Space Council and the astronaut-dominated NASA senior management. Richard Truly was not keen on such big programs given NASA's struggles with the space shuttle and the International Space Station. Despite this Quayle took the proposals to Bush, who decided to pursue both the lunar and Mars goals simultaneously. In response NASA produced a preliminary estimate of the costs, \$400 billion over 30 years. This sum included many unnecessary items that could be discarded, although the report did not say that. It was leaked to the press, the politics got nasty and the project collapsed. Richard Truly was fired and the National Space Council folded.

The next president, Bill Clinton, was not a great fan of space. After he left office, he produced a 100,000-word memoir that hardly mentioned space exploration at all. During the Clinton administration, space shuttle flights continued, and the construction of the International Space Station began.

The loss of space shuttle Columbia and its crew early in George W. Bush's presidency changed everything.

GODDESS OF
THE MOON

———————•———————

In the space age so far four crews have perished during a
mission. Vladimir Komarov died when his Soyuz return
capsule failed, and he plunged into the ground in 1967. The
three-man crew of Soyuz 11 became the only people to have
died in space when their capsule depressurised in 1971. Space
shuttle Challenger exploded after lift-off in 1986, and then
there was space shuttle Columbia. It will happen again. As
we start being more adventurous, we will mourn the loss of
a crew more than a few times in the next 50 years in space.

Columbia began its 28th return to Earth after sixteen
days in space completing the STS-107 mission on 1 February
2003. It had been a wide-ranging science mission studying
the atmosphere and microgravity. As they prepared to return
to Earth, they did not know the peril they were in, but on
Earth many suspected they were doomed.

About 82 seconds after launch, at an altitude of
20,000 metres, a suitcase-sized piece of thermal insulation

foam had broken off the external fuel tank and struck the leading edge of Columbia's left wing, creating a 10-centimetre hole. The same had occurred, to a lesser and non-fatal extent, on at least four previous shuttle flights.

Aware that something had happened, controllers looked at video taken of the launch and concluded that it was nothing unusual. But the day after, higher-resolution video revealed that the foam had struck the left wing; the extent of the damage, though, was not possible to ascertain. Engineers wanted to use top-secret spy satellites to look at Columbia, but officials did not think it was necessary. Another engineer requested that an astronaut visually inspect the area, but he was turned down. The subsequent accident report said that the engineers found themselves in the position of having to prove that the shuttle was unsafe – a reversal of the usual procedure.

At 2.30 a.m. ET on that awful morning, the Entry Flight Control Team began their shift at Houston's Mission Control Center. It was a normal re-entry. The weather at the Kennedy Space Center landing site in Florida was good. At 8.00 a.m. Flight Director LeRoy Cain polled mission control for a go/no go. A few minutes later the capcom (capsule communicator) notified the crew that they were go for the de-orbit burn. Travelling upside-down, tail first over the Indian Ocean, they were on their 255th orbit. The two-minute 38-second burn slowed them to start their entry into the atmosphere. There was no turning back.

Commander Rick Husband turned Columbia around and pitched its nose up, allowing the heat-resistant tiles on its belly to face the heat of re-entry friction. At 8.44 Columbia reached 121,000 metres – the so-called entry interface when the first signs of the atmosphere are evident. They were over the Pacific Ocean.

Four minutes later a sensor on the left wing's leading edge showed strains higher than those seen on previous re-entries, but the data was stored on the onboard recorder and not sent to ground controllers. Travelling at Mach 24.5, Columbia made a planned turn to the right. Then began the ten-minute-long period of peak heating. They were nearing the Californian coastline. Wing leading edge temperatures were approaching 1,450°C. Altitude was 70,500 metres.

People reported seeing debris from Columbia. Four hydraulic sensors in the left wing went 'off-scale low', indicating they had failed. Witnesses reported seeing a series of bright explosions. At 8.59 and 15 seconds pressure readings had been lost from both left-wing landing tyres. Rick Husband had been trying to say something. Seventeen seconds later he said, 'Roger, uh, bu—'. He was cut off. Columbia was breaking up. Seven minutes later, flight controllers knew the crew had died.

The cabin remained intact as Columbia broke apart. Rick Husband would have worked the hand controller, trying to regain control, but the tumbling would have indicated to all that there was no shuttle left. We do not know how long they survived before the cabin disintegrated about 30 seconds later; certainly the crew understood their fate. The investigation concluded that they died due to 'blunt trauma and hypoxia with no evidence of lethal injury from thermal effects'. Three of the crew were not wearing gloves at the time of the accident and one was not wearing a helmet.

A year later President George W. Bush unveiled plans for what he called a sustained and affordable space exploration program, beginning with returning astronauts to the Moon by 2020. NASA began developing a new manned spacecraft – the Crew Exploration Vehicle, later renamed Orion – and the Ares rocket, the world's most powerful. Bush called it the Vision for Space Exploration. There was one crucial condition. The space shuttle had to retire. NASA's Administrator Michael Griffin was keen on the plan and began a study called the Exploration Systems Architecture Study (ESAS), looking at the details. The project became known as Constellation. It could have worked but Congress sliced some money from its request, so the project would do less and take longer, and next year some more money was removed. Eventually it was clear that the project was going nowhere with the resources it had left.

When he came into power, President Obama changed everything. If you could summarise his plans it would be: forget the Moon, send people to Mars instead – but not to land, just orbit and return. The Mars trip he envisioned for 2035. The centrepiece of his crewed space program was a mission to an asteroid that would probably take about two months if a suitable near-Earth passing asteroid could be found. Constellation was dead, although the Orion capsule program survived, and the House of Congress took over the big rocket which became the SLS – the Space Launch System. NASA later modified the plan to robotically bring a piece of an asteroid into lunar orbit and send astronauts there to examine it.

Moon supporters and many Apollo astronauts were not happy with Obama's plan to bypass the Moon altogether. In May 2010 Neil Armstrong went before a Senate hearing to question Obama's new vision. He said that he worried about the possibility that NASA would lose its edge in spaceflight: 'If the leadership we have acquired through our investment is simply allowed to fade away, other nations will surely step in where we have faltered ... I do not believe that this would be in our best interests.' Armstrong, along with Eugene Cernan, the last astronaut on the Moon, said that the Obama plan was short on ambition. Cernan said that he, Armstrong and Apollo 13 Commander James Lovell agreed that the administration's budget for human space exploration 'presents no challenges, has no focus, and in fact is a blueprint for a mission to nowhere'. Buzz Aldrin, Armstrong's partner in the Apollo 11 mission supported the president's plan. At its unveiling at the Kennedy Space Center in April 2010 President Obama said of the Moon, 'We've been there before. Buzz has been there.'

Armstrong said the Obama plan was 'contrived by a very small group in secret' who persuaded the president that this was the way to put his stamp on the space program. 'I believe the president was poorly advised,' he said, in what many regarded as a criticism of Aldrin. Cernan was even harsher, saying Obama's space budget projects either show extreme naivety or a willingness to accept a 'plan to dismantle America's leadership in the world of human space exploration'. John Holdren of NASA countered by saying the plan was based on findings of the US Human Space Flight Plans Committee, also known as the Augustine Committee, which had six public meetings. Part of the Obama plan was for more commercial involvement in space. He wanted private industry to take over putting astronauts into space, considering that it might take the burden off government spending. Republican Senator Kay Bailey Hutchison of Texas said to NASA Administrator Charles Bolden, 'You are putting all of our dreams and hopes and taxpayer dollars into this commercial investment.' Bolden said the target was to have the first crewed commercial flight to the space station by 2015. It didn't happen.

The next president, Donald Trump, in time-honoured fashion, threw out his predecessor's plans almost completely. On 30 June 2017, he signed an executive order to re-establish the National Space Council, chaired by Vice-President Mike Pence. The Trump administration initially kept Obama's spaceflight programs in place: commercial spacecraft to ferry astronauts to and from the International Space Station, the government-owned Space Launch System, and the Orion crew capsule. Trump, however, wanted to reduce Earth science research and called for the elimination of NASA's education office.

Later that year in December Trump changed the direction of space exploration completely with his Space Policy Directive 1 and its goal of a 'US-led, integrated program with private sector partners for a human return to the Moon, followed by missions to Mars and beyond'. It called for NASA to 'lead an innovative and sustainable program of exploration with commercial and international partners to enable human expansion across the solar system and to bring back to Earth new knowledge and opportunities'. Trump said, 'The directive I am signing today will refocus America's space program on human exploration and discovery ... it marks a first step in returning American astronauts to the Moon for the first time since 1972, for long-term exploration and use. This time, we will not only plant our flag and leave our footprints – we will establish a foundation for an eventual mission to Mars, and perhaps someday, to many worlds beyond.'

NASA was aiming for a crewed landing on the Moon by 2028 but in March 2019 Trump upped the ante. Do it sooner, he said. So Vice-President Mike Pence instructed NASA to put astronauts on the lunar surface by 2024, four years earlier than previously planned. He said such urgency was required to safeguard the country's leadership and dominance in space. '2028, that's just not good enough,' Pence said during the fifth meeting of the National Space Council (NSC), which he chaired. 'We're better than that.'

Following Trump's abandoning of Obama's plan, NASA seamlessly repurposed the asteroid rendezvous mission into a power and propulsion module for the Lunar Gateway we encountered earlier. This small space station will be placed into a highly elliptical seven-day so-called near-rectilinear halo orbit around the Moon, which would bring the station within 3,000 kilometres of the lunar north pole at closest approach and as far away as 70,000 kilometres over the south pole. It would take five days to get there from Earth and five days to get back. A crew of four could live there for eleven days, which is determined by the capabilities of the Orion spacecraft.

Led by NASA, it will have international partners and major commercial involvement. After the first few lunar landing missions, which will enter lunar orbit before landing, it will ultimately be the place astronauts travel to before setting off to land on the Moon's surface and a proposed starting point for spacecraft going to Mars.

Some people do not like the Gateway idea. Veteran Mars advocate Robert Zubrin said, 'This boondoggle will cost several tens of billions of dollars, at the least, and serve no useful purpose whatsoever. We do not need a lunar orbiting station to go to the Moon. We do not need such a station to go to Mars. We do not need it to go to near-Earth asteroids. We do not need it at all. If we do waste our time and money building it, we won't go anywhere.'

The program to return astronauts to the Moon was given the name Artemis: the twin sister of Apollo, the goddess of the hunt, the wilderness and the Moon. According to current plans Artemis 1 will set off from the Kennedy Space Center Launch Complex 39B in 2020 or 2021. An unmanned Orion spacecraft will be sent on a mission of 25½ days, six of them in lunar orbit. The mission will certify the Orion spacecraft and Space Launch System rocket for crewed flights.

The Space Launch System in the Artemis 1 configuration will have a first stage consisting of a core of four rocket engines of the type used in the space shuttle, augmented by two strap-on rockets powered by solid fuel, again similar to the shuttle. The upper stage will have a single rocket motor whose design is based on the motor used in Delta rockets, which have an excellent track record. The Orion capsule will be connected to a European-built service module, repeating the Apollo spacecraft configuration.

Onboard will be many instruments, in particular one to measure radiation, which is vital to understand for the safety of subsequent crews. A radiation protection vest called AstroRad will be evaluated for the first time out of low Earth orbit. In the future it will be a vital piece of protection. Two female mannequins will occupy the seats, one wearing AstroRad and the other not. A series of small satellites, called CubeSats, will be deployed during the flight. One will measure how yeast is affected by radiation, another the particles and magnetic fields that stream from the Sun. Several will look for ice on the Moon.

Artemis 2 is currently set for September 2022 and will carry a crew of four around the Moon on a week-long mission. It will be the first time humans will have left low Earth orbit since Apollo 17 in 1972. The first time it had been done was Apollo 8 in December 1968. Back then, for such a dramatic moment the execution was very subdued. The person talking to the crew was Michael Collins who was part of the Apollo 11 crew. He said, 'Apollo 8. You are go for TLI [trans-lunar injection].' Mission commander replied, 'We understand. We are go for TLI.' And that was it. Humans leaving the vicinity of Earth for the first time. No drama.

Artemis 2 will orbit Earth twice while periodically firing its engines to build up enough velocity to push it towards the

Moon before looping back to Earth. It will take place prior to the assembly of the Lunar Gateway in lunar orbit, which will occur between 2022 and 2023.

The climax to this series of missions – the Artemis 3 mission – is slated for 2024: the first landing on the Moon since Apollo times.

THE SUBTLE ACCOMPLICE

———•———

Where are we going to? Most people do not know the Moon. The Earth they know, easily picking out Europe and Africa and all the other continents. But the Moon, the names of the major craters and 'seas' are a mystery to many. At the time of Apollo many offices and children's bedrooms had a poster of the Moon. They are seldom seen now. Perhaps we will all see more of them in years to come.

Unlike the Sea of Tranquillity where Apollo 11 landed near the Moon's equator, it is very difficult to see Artemis 3's landing zone from Earth. Even under the most favourable conditions it is only a thin sliver of light seen at the very edge of the Moon. It lies in the ancient, rugged southern uplands and to find it through a telescope you have to hop from one crater to another, starting at Tycho with its impressive ray system. Then your eye moves southwards to the magnificent crater Clavius, one of the Moon's oldest formations, with its picturesque curving chain of internal craters. Thence over the terraced ramparts of Morteus and beyond the dark shadows of Cabeus.

During the return to the Moon, Artemis 3 will descend under power towards the Moon's shadowy south polar zone with the Sun beaming into its cabin from the left. Nearing the landing site, the Artemis astronauts will arrive at Mount Malapert, five kilometres high and with its peak almost always in sunlight. They are now in the final stage of the landing, within two craters of their destination.

Rising out of a vast shadow zone before them is the bright rim of the crater Shackleton that marks the Moon's south pole. Flying over the dark nothingness that is the depth of Shackleton crater, they head for its illuminated far rim. The beacon and other pieces of equipment already positioned there are flashing their red lights as they begin the terminal phase of the landing 400 metres above the primary landing zone. The astronauts slow their forward motion, look carefully at the ground below, reducing thrust, kicking up some dust as they touch down, 52 years after the last time. The best part of a lifetime for them, but to the Moon, nothing.

Neil Armstrong said: 'There are secrets on the Moon, there are things to see beyond belief.' The Moon has always been with us, its gravity and its changing light affecting all beneath it. There are some linguists who believe the origin of the word 'men' comes from an earlier word for 'moon'. Every creature that ever lived has done so under the Moon. Its influence has been seen as beneficial yet somehow unsettling, an unseen power that was darker and more mysterious than the unsubtle Sun. Its life in night-time skies meant that it was not just an observer of human affairs but a participant in them, as well as an accomplice to the strange forces of the night. On its ancient surface is a record of the solar system's life, a chronicle long since obliterated from the face of the Earth. It is a history we have learned to read. 'What is there in thee, Moon! that thou shouldst move my heart so potently?' wrote Keats. Very few people know the Moon. It's not taught in schools.

Seen from the Earth when the Moon is a crescent, craters appear as dark elongated streaks and the border between day and night – the terminator – is delicately ragged by the Moon's unevenness. The craters are not really as deep and steep-sided as they seem. Even the most spectacular ones rarely have slopes that exceed 40° and most of them are much shallower. If you stood inside one of these great walled plains its ramparts would be far over the horizon. The Moon is much smaller than the Earth so its surface curves away more sharply. No human or robot has ever seen the Sun rise and set from the lunar surface.

Early in the revealing of the Moon as seen from the Earth that takes place every month by the rising Sun is the magnificent Mare Crisium – the Sea of Crises. To me few sights on the Moon are as spectacular as watching the Sun illuminate the mountains that surround this vast oval-shaped, lava-filled basin. Nearly 4 billion years ago, when the solar system was young and still strewn with rocky debris, something large, by which I mean perhaps a few tens of kilometres in size, slammed into the Moon here. The explosion scooped out a depression almost 600 kilometres across, scattered debris all over the Moon and sent shock waves pulsing through it. Once, Mare Crisium was a real sea, of molten rock. Between

300 and 500 million years after the impact, lava seeped up through the lunar crust and filled the basin, drowning some craters, partially obscuring others. There are lava flows superimposed upon flows. Crisium is one of the smaller of the lunar 'seas', but one of the most interesting, and certainly, by grazing light, one of the most beautiful.

From Crisium three Soviet missions – Luna 16, Luna 20, and Luna 24 – collected and returned pieces of the lunar surface after several failures in the same area. Scientists using the camera aboard the 2009-launched Lunar Reconnaissance Orbiter have been able to see their remains. Under a low light numerous wrinkles or ridges, at most only a few hundred metres in height, can be seen. They are termed dorsa and they cast long shadows. Follow one of them, the Dorsa Harker, south and it will lead you to the site of Luna 24, the last Moon lander and the third in the Soviet Union's automated sample return missions. Not far away lies the damaged Luna 23 craft, designed to carry out the same task but crippled during a very rough landing. Its sample return canister may still be attached forlornly to the main body of the craft, or the rising Sun may come across it piece by piece scattered over the lunar plain.

If you are lucky and catch the Moon at just the right time, and there are astronomers who spend their lives waiting for the correct lighting conditions, you may see the peaks of Crisium's eastern wall greet the lunar dawn. For me this is one of the most thrilling moments of Moon-watching. For an hour or two the peaks appear as a strange constellation of stars, the myriad points of light growing as sunlight races down the slopes. Then comes a moment, perhaps my favourite moment of the whole lunation, when the Sun pokes its head above the rim and stretches across the Mare. Sunlight pours into the dark, throwing the wrinkles and ridges in the Mare's surface into high relief.

One day, perhaps, in the far future when there are many bases on the Moon, explorers might drive their pressurised rover from Crisium Base to follow another wrinkle ridge, the Dorsum Termier, south, on a route that will bring them to the wreckage of Luna 15, one of the strangest events in the exploration of the Moon. As Apollo 11 was en route for the first lunar landing, a Soviet craft was already in lunar orbit. It was an attempt to steal the thunder of the first manned landing. Just a few hours after Armstrong and Aldrin touched down on the Sea of Tranquillity, Luna 15 was coming in to land on the Sea of Crises.

The brightest crater in the vicinity is Proclus, the second brightest object on the Moon after Aristarchus. It was formed only about 50 million years ago when an asteroid, approaching at a shallow angle from the southwest, slammed into the Moon and scooped up bright highlands material which it scattered across the surface in a series of brilliant arcs. From orbit, Mare Crisium and Proclus are even more spectacular than through a telescope. Only a few people have seen them 'up close and personal' and one of them was Apollo 15 command module pilot Alfred Worden. He said: 'It is almost like flying above a haze layer and looking down through the haze at the surface. Ejecta from that crater [Proclus] does not look like it is resting on top of Crisium. It looks like it is suspended over it.'

Look south at the Sun rising over the ramparts of Langrenus, a 136-kilometre crater on the eastern shore of Mare Fecunditatis, the Sea of Fertility. The crater's name is in honour of Dutch mapmaker Michel van Langren, who in 1645 drew the first map of the Moon. In 1992 the veteran lunar observer Audouin Dollfus of the Observatoire de Paris was observing Langrenus when he noticed a series of changing glows. It may have been gas vented from the extensive series of fractures that can be seen on the floor of the crater. The Moon, it seems, is not such a dead world as was supposed. One of the places where automated observation stations could be placed is in Langrenus.

Only a few hours after Mare Crisium, the Sea of Fertility emerges from the lunar night, larger than Crisium but without its well-defined borders: towards the south it peters out into the highlands near the crater Petavus. To the north a small patch of highland provides the border with Mare Tranquillitatis, the Sea of Tranquillity. From Langrenus you can take your telescope into the southern uplands. Here the terminator is constantly changing as it moves across uncountable craters.

When I look at the Moon through my telescope, which is not as often as I would like, my gaze is always drawn to a particular area if it is in sunlight. It's a region that borders the eastern shore of the Sea of Serenity, on a tongue of lunar highland that separates Serenity from Tranquillity. Of all the places where man touched down on the Moon, the general region of the Apollo 17 landing site is the easiest to see through a telescope. The exact valley, though, is elusive: it needs a high power and good atmospheric conditions, but when you catch it you get a special thrill. This is Taurus-Littrow. Six hundred and fifty kilometres away to the south-west, across a vast and largely featureless lunar plain, is the first lunar landing site, occupied a mere three-and-a-half years earlier.

Was it only for just three-and-a-half years, and so long ago, that we walked on the Moon? There was serenity, tranquillity and sadness in the voice of Gene Cernan, the last man on the Moon. At the end of his final moonwalk he took the TV camera and pointed it at his spacecraft's legs. There was a plaque with words that sounded so final. He removed the cover and read the message: 'Here man completed his first explorations of the Moon, December 1972.' He added: 'This is our commemoration that will be here until someone like us, until some of you who are out there, who are the promise of the future, come back to read it again.' As he spoke, the Earth was high in the south-western lunar sky.

Sometimes I look for the sharply defined crater Taruntius on the area of highland between Serenity and Tranquillity. As Armstrong and Aldrin began their descent to the lunar surface for the first lunar landing in July 1969 they looked down on this region and searched for familiar landmarks to guide them on their descent across the south-eastern shores of Tranquillity until they landed near the lunar equator.

On day five as viewed from Earth, the magnificent crater Theophilus can be seen where Mare Nectaris meets Mare Tranquillitatis. Theophilus is one of the Moon's finest craters with a grand central peak complex. Nearby are the equally magnificent Cyrillus and the dramatic Altai Scarp, formed at the same time as the impact that gouged out the Nectaris basin. The Apollo 16 mission collected several pieces of basalt believed to be ejecta from the formation of Theophilus.

By now the broad crescent of the Moon is visible during the afternoon and for most of the evening. It will transit, or pass across the meridian, around sunset, so the best time to observe it is during the first hours of darkness. The whole of the Sea of Tranquillity – the Moon's most heard-of region – is now visible. In the south-west of Tranquillity lies the most famous of the many human artefacts left on the Moon: the base section of the Apollo 11 landing module, Eagle, from which Neil Armstrong made the 'small step' in 1969. Nearby are three small craters now called Armstrong, Aldrin and Collins after the crew. Normally lunar features are named posthumously; these are the only nearside features that break that convention. Only 40 kilometres to the north lies the Surveyor 5 craft that soft-landed in September 1967. Turn north-east and travel a further 60 kilometres and you will reach the place where Ranger 8 struck the Moon in February 1965 and formed a small crater as it was vaporised.

The Sun now rises on Apollo 16, the only manned landing site in the highlands and away from the Maria. It is the most difficult of the six landing sites to locate through a telescope. You have to crater-hop westwards from Theophilus to identify the spot but when you do you can actually see the splash pattern of South Ray crater. The astronauts trampled over this region and the material they brought back was found to be only 2 million years old, as opposed to other Apollo samples that were billions of years old.

Look towards the centre of the Moon. Adjoining the Sea of Vapours to the south-west is Sinus Medii or the Central Bay, the nominal centre point of the lunar disc from which latitude and longitude are measured. Two United States Surveyor craft, numbers 4 and 6, landed within a few kilometres of each other in Sinus Medii. Contact with Surveyor 4 was lost but it may have landed safely. Surveyor 6 was successful and came to rest near a small ridge. You could walk from one to the other in less than an hour.

Now comes the hour that everyone who knows the Moon has been waiting for. The eastern end of the enormous and majestic Mare Imbrium, the Sea of Rains, is glimpsed for the first time. So vast is this sea that it will take a further three Earth nights for it to emerge. Someday explorers will visit the grave of Luna 2 that lies hereabouts. It was the first man-made object to touch the Moon, over sixty years ago. But, despite Soviet propaganda at the time, there is no scattered field of metal Soviet pennants. The impact would have been so great that what would be seen would be a small crater and probably no trace of metal in the vicinity. That would have been vaporised.

In 1971, Apollo 15 astronaut David Scott saw a small white rock which he recognised as anorthosite and thought to be part of the Moon's primordial crust, the so-called genesis rock. It turned out to be a misnomer as when it was analysed back on Earth it was found that the rock was younger than the crust. Apollo 15's remains lie not far from here in the most dramatic of the manned landing sites. Hadley Rille, a winding gash across the surface that is probably a collapsed tube that once held lava, lies in shadow for a while after local sunrise. Look on photographs or with a strained eye at the eyepiece and if you are lucky you can see the elbow where astronauts Scott and Irwin looked into the chasm.

Look to the south and you may now see Tycho. After Surveyor 6 landed on the Moon in 1967 the Surveyor mission goals had been achieved so Surveyor 7 was given to the scientists to land just where they wanted, and they chose Tycho, the source of the Moon's greatest ray system – debris scattered all over the Moon by the impact that formed the crater.

By day nine of the Moon's lunation cycle the terminator has reached one of the Moon's finest sights: Copernicus. Although not the largest crater (that title belongs to Bailly in the southern uplands), nor the deepest (probably Newton), nor the brightest (Aristarchus) nor the one with the largest ray system (Tycho), it is undoubtedly one of the most dramatic features on the Moon. In 1966 the United States Lunar Orbiter 2 was only 45 kilometres above the Moon and looked northwards across Copernicus. The image it took was called the 'picture of the century': in it, you can see the ramparts and the plains in between and the 400-metre central peak. Before it was cancelled, Apollo 20 was due to land on that central mountain and sample material from inside the Moon. Look at it now and wonder what could have been.

To the north of Copernicus, the whole of Mare Imbrium is seen, a smooth ellipse bounded by mountain ranges: the Alps, the Caucasus, the Apennines and the Carpathian

Mountains. On the north-western edge of Imbrium lies the beautiful Sinus Iridium, or the Bay of Rainbows. Ringed by the Jura Mountains this is one of the most enchanting sights on the entire Moon. At sunrise when the Jura Mountains are illuminated they seem to protrude into the blackness and give a jewelled-handle effect. Someday, someone will be the first to stand in this bay, on the almost black lunar surface, the stars above and in the west a wavy, horizontal sunlit band of lunar mountains; in the east: the glow of the solar corona and the zodiacal light that herald the dawn. Nearby lies the silent Lunokhod 1. For more than a year starting in 1970 it was the first wheeled vehicle on the Moon. Remotely controlled from the Earth, it roamed across 11 kilometres and saw over 200 lunar panoramas. It survived several lunar nights.

The Sun now reaches the landing site of Apollo 14 at Fra Mauro, a region of battered craters with broken rims inundated with lava. With my telescope I can pick out the site. Here is material ejected from the great Imbrium basin 600 kilometres to the north. Illuminated shortly after-wards is Apollo 12's landing site. It descended onto Oceanus Procellarum – the Ocean of Storms – and when Pete Conrad looked into the mid-distance, 180 metres away, on the slope of an ancient crater, was Surveyor 3, which had landed over two years earlier.

By now the crater Gassendi is almost fully illuminated. Nearby craters Grimaldi, Riccioli and Hevelius, are named after early mappers of the Moon. For me Grimaldi stands out as conspicuously dark. Here another cluster of spacecraft lie scattered across the lunar plains. In December 1965 the

Soviet Luna 8 crashed here, whereas in the following year Luna 9 made the first soft landing on the Moon and sent back the first picture from the lunar surface. Luna 13 also lies hereabouts. It was the third spacecraft to successfully land on the Moon. On the edge of Procellarum, not far from Luna 9, is the tiny light speck of the crater called Galileo, hardly a fitting tribute to the man who first saw the Moon for what it is: a mountainous, three-dimensional world.

Soon Mare Serenitatis is lost to the darkness and the Sun sets on the Caucasus Mountains on its north-west edge and the Apennines that border Mare Imbrium, which is soon to go into shadow. To my mind, at this time the lengthening shadows make the Apennines look especially three-dimensional. The late phases of the Moon have an emotional power because they are not often seen, as the Moon rises after midnight. Despite the beauty of its waning crescent there is somehow a feeling of loss when you observe the Moon at this time. Soon it will only be visible for a short while before sunrise, low in the east, and then the Moon is gone.

Gone from the skies for three days until it returns as a crescent, but not gone from our minds. The Moon's long isolation, only briefly interrupted 50 years ago, is about to end forever.

DARKNESS AND LIGHT

———•———

The Moon's axis of rotation, unlike that of the Earth, is not markedly tilted compared to its orbit around the Sun. This means that the Moon does not have strongly different seasons and that the Sun never rises very far up in the polar skies. You could stand at the lunar poles for a whole year and the Sun would only move up or down about 1.5 degrees. Imagine a place of long shadows that every month point in all compass directions. Some of the shadows are present all year round. The Sun skirts the horizon, as does the Earth. Sometimes they are seen together, sometimes one or the other dips below a distant mountain peak.

Because of the low-angle Sun, many crater floors or areas surrounding high relief will be in permanent shadow. According to Ben Bussey of NASA, 'The floors of these craters never see the Sun and are therefore extremely cold; some are only 25 degrees above absolute zero. They are cold enough that any water molecule from a cometary impact that enters them will not have enough energy to escape.' It's colder in the depths of these craters than it is on the surface of distant Pluto.

Topographic maps show that the north polar area displays relatively little relief whereas the south polar region has a much more rugged terrain. In the north the permanently shadowed regions may be concentrated at the bottom of large craters, forming a crescent in the wall and floor areas that are furthest from the pole. Crater Hermite, at 86° N, may be one such crater. Several small lunar craters on the floor of Peary and three larger craters aligned roughly along the 315° longitude line are also protected from sunlight. These five regions constitute the largest potential deposits at the north pole.

In contrast, according to Erwan Mazarico of NASA's Goddard Space Flight Center, 'The south contains more large craters, and the difference between the deepest and the highest locations near the South Pole is twice what you find near the North Pole. In terms of receiving sunlight the South hosts the most and best high-illumination areas, in part thanks to an elevated mound connecting the Shackleton and de Gerlache craters.'

Those high-illumination areas are the other key to the importance of the lunar poles, and Shackleton crater in particular. Somewhere in the Moon's polar regions, it was once thought, as well as the eternal dark there might be a mountain peak that catches the Sun's rays all year round. A 'Peak of Eternal Light'. It would be a great boon for a Moon base to have continuous access to solar power. It's only possible at the poles as the rest of the Moon experiences two weeks of daylight followed by two weeks of darkness during which power would have to be supplied by batteries or nuclear energy.

So is there a Peak of Eternal Light? There are two ways to look for it. One is to collect observations and see if a particular peak is ever in shadow; the other is to use data from orbiting spacecraft that have used radar and laser pulses to map the lunar topography. According to Erwan Mazarico, 'We surveyed both polar caps with the best topography models of the lunar poles, which were obtained by the Lunar Orbiter Laser Altimeter (LOLA) instrument onboard the LRO. With about 6 billion topographic measurements globally, the accuracy and resolution were sufficient to precisely calculate what the local horizon is from any point near the lunar poles, and then run long-duration simulations of solar illumination. Then it's easy to find the locations that are illuminated the most, and to know how long the "night" would be at that location.'

Alas, detailed observations revealed there is no Peak of Eternal Light. 'No place on the lunar surface receives constant illumination,' says Mazarico. 'The best location is on Shackleton's rim where there is less than a square kilometre which is in view of the Sun 92.5 per cent of the time, but we found that just 10 metres above the surface [to avoid very local obstacles] that percentage rises to 95.8 per cent. This means that by placing solar panels above surface level by only 10 metres one could increase the duration of continuous sunlight to 240 Earth days and reduce the duration of the longest "night" to just 36 hours.'

So, there is a part of the Moon that has, within walking distance, regions that are in almost permanent sunlight and regions that are in permanent darkness. This then, not the equator or anywhere else, is the place to put a base on the Moon. Electricity-producing solar arrays could be placed in the bright areas and connected by a microwave or cable link to the habitation region, producing constant energy. The dark areas could be mined for ice.

Another interesting fact is that the south pole thermal environment is less harsh than that found at the lunar equator. There are few dramatic temperature shifts. Surface temperature remains close to −30°C. Outside the polar regions surface temperature spans about 400° over the course of the 28-terrestrial-day lunar day. The fourteen days of hot sunlight and fourteen days of frigid darkness does not occur at the poles. Instead, the Sun is on the horizon for most of the lunar day. This will mean designing a lunar base will be easier here.

PASSPORT TO
THE PLANETS

———•———

On Saturday 31 July 1999, 30 years after Apollo 11, thousands of telescopes all over the world were pointed towards the Moon's south pole. They were all waiting for the finale of a spacecraft that had been circling the Moon for eighteen months. It was to crash into a permanently shadowed crater.

During its life the Lunar Prospector spacecraft had changed what we knew of the Moon. It had produced global maps of the Moon's gravitational and magnetic fields and the distribution of its key elements, and had given us a much better understanding of its origin, evolution and composition. It might also have discovered the Moon's most precious resource, water ice, which is vital for a colony and from which can be produced liquid hydrogen and liquid oxygen to be used as rocket fuel. In its final act Lunar Prospector was to strike a region and hopefully throw some of the surface material – possibly including ice – up into space where it could be seen. Few expected it to work, but it was worth trying. It was launched in January 1998 and entered a lunar orbit that took it regularly over the poles as its five science instruments began their survey. It was part of a pair of space probes that ended many years of neglect of the Moon. Together they showed us there was much more to be discovered on this seemingly dead world. Before Lunar Prospector we didn't know the real Moon.

The first of these probes, Clementine, circled the Moon for a while in 1994. It was a military mission evaluating space technologies without violating any then treaty restrictions, but astronomers asked if they could do lunar science with it without interfering with its main purpose. The probe did not carry instruments designed to look for lunar ice, which had long been speculated about but for which there was scant observational evidence; however, its controllers improvised an experiment based on radio waves being reflected from the lunar surface differently depending on its composition. Radio waves are scattered in all directions from surfaces made up of ground-up rock like those found on most of the Moon, Mercury, Venus and Mars but they are reflected more strongly from icy surfaces, which act like a mirror. When the dark regions of the Moon's south pole were surveyed, the data returned suggested there may be deposits of ice there, perhaps as much as a billion cubic metres – the volume of a small lake. It was suggestive but not conclusive.

Lunar Prospector had taken just 22 months to build and was less than 2 metres in diameter and 1.5 metres long. It had sensors on deployable arms and six thrusters controlled the spacecraft's direction in space. To save money it had no onboard computer, it just sent the data from its sensors straight back to Earth. The hopes of finding ice on the Moon rested with the Los Alamos Laboratory's neutron spectrometer. If it is there within about one metre of the surface, that sensor stood a good chance of finding it. The spectrometer needed the help of particles that fly through space from other parts of the galaxy and beyond. These high-energy cosmic particles slam into the Moon's surface throwing up neutrons. If hydrogen, a component of ice, is present then some of the neutrons would be slowed down. The spectrometer was designed to capture a sample of those neutrons and measure their energies to look for the low-speed neutrons that were the signature of water. The researchers said that they would know within a month or two. One said, 'If we see water, I suspect the land rush is on.'

After three weeks rumours began circulating that Prospector had detected slow neutrons as it passed over the lunar poles. The official announcement came on 5 March 1998. Prospector had indeed found ice. The following day newspapers all over the world said it was the biggest breakthrough in space exploration since Neil Armstrong stepped onto the Moon's surface.

Lunar Prospector had found between 10 million and 300 million tonnes of water, not only as expected in the dark depths of the south pole but also in the craters of the north as well – in fact there was 50 per cent more water in the craters of the north than those of the south. Lunar Prospector's chief scientist Alan Binder said, 'We have the first unquestionable results indicating that there are significant quantities of water at both lunar poles.' The view was that the water was in the form of ice crystals at a concentration of about 1 per cent by volume of the lunar soil. Three hundred million metric tonnes, enough to form a ten-metre-deep lake of ten square kilometres. 'The implications are tremendous. For the first time we can go to a planetary body and we can fuel up. That fuel can be used to go to Mars and anywhere in the solar system,' said Dr Binder. He added that if you picked up a cubic yard of lunar soil from the ice field then you may find one, two, maybe five gallons of water.

The Times of London said, 'Galileo's vision of watery moon is proved right'. The *Daily Mail* gave over the full front page: 'Man may live on moon within 30 years.' The associated article said, 'The moon could become a vast launching pad for missions to Mars and the rest of the solar system.' The lead story in the *Daily Telegraph* was 'Water found on moon boosts space travel hopes'. The *Express* said, 'Next stop planet Mars.' On its front page was a banner across the top and a picture of the lunar limb: 'Could your children be living here soon?' In the *Guardian*: 'Water on the moon – our passport to the planets.'

So, the evidence for the ice rested on two lines of evidence. The Clementine radar measurements show the unusual radar signature only in the permanently shadowed regions near the poles and Prospector's slow neutrons are only detected over those same regions. The case is not absolutely conclusive but is very good. To prove it we will have to go there.

Knowing it had already changed the Moon as we saw it, astronomers awaited the final few minutes of Lunar Prospector's life. The estimated time of the impact passed. What had happened 400,000 kilometres away? There were no signs of water. There could be many explanations: the spacecraft might have missed the target area, or it might have hit a rock or dry soil, or there may be no ice at all. After looking at Prospector's final transmissions NASA said it landed on target, but even given the most favourable circumstances, the dust plume would have stretched only 22 kilometres above the Moon's surface and would have been 100,000 times fainter than the brightness of the lunar limb. It was then realised that the ice debris plume would never have been detectable from Earth.

But as well as looking for ice Lunar Prospector did something else as it plunged onto the Moon's surface. It also buried someone on the Moon. A secondary mission for the spacecraft was to deposit one ounce of the cremated remains of Eugene Shoemaker, making him the first person to be buried on another world. The renowned geologist had been killed in a car crash two years earlier. Shoemaker was a legend among geologists. Almost on his own he invented the science of the study of cosmic impacts and he played a key role in training the Apollo moonwalkers to explore the Moon in a scientific manner. His work meant the scientific return from Apollo was remarkable. He had wanted to be an astronaut himself and perhaps today he could have been. But in the early 1960s health qualifications were more stringent than they are now, and he was turned down because of a minor medical problem. Shortly before Professor Shoemaker died, he said, 'Not going to the Moon and banging on it with my own hammer has been the biggest disappointment in life.' In death he got his wish. 'I don't think Gene ever dreamed his ashes would go to the Moon,' his widow Carolyn Shoemaker said shortly before watching Lunar Prospector blast-off. 'He would be thrilled. We will always know when we look at the moon, that Gene is there.'

It was thought that the discovery of ice would give a boost to planned lunar missions, including a European concept funded by private industry. But over the subsequent years it seemed that governments and space agencies were lukewarm on the idea of going back to confirm the discovery.

At the time the European Space Agency (ESA) had a concept for a soft lander called LEDA and an orbiter called MORO, but both were shelved because they were too expensive and lacked political support. But the ice discovery did add some impetus to a concept being championed by a former astronaut. EUROMOON 2000 was the brainchild of Wubbo Ockels, who flew on space shuttle Challenger in 1985. He had spent seven years preparing the proposal.

At the time the space science interests of ESA were dominated by astronomers who had little interest in the Moon. So Ockels took his case to European industry, carting a three-dimensional model of the lunar south pole into the offices of space executives, using a flashlight to demonstrate why parts of it were in eternal darkness and to show the regions where a spacecraft could land. It was a strategy that worked. He gained three-quarters of the $200 million cost from

the private sector. Aerospace and high technology companies as well as other sponsors wanted to be associated with
a high-profile return to the Moon. It was unprecedented for
a mission to be so well supported by non-government funds.
The director general of ESA, Antonio Rodotà, wrote to the
head of NASA, Dan Goldin, suggesting they cooperate on
the mission.

The mission would involve a small orbiter called Lunarsat
to scout the south pole for landing sites. The following year
a lander would be launched to the rim of Shackleton crater
to begin setting up a 'robotic village' for scientific studies
and technology demonstrations. The mission was particularly challenging because it required a landing accuracy of
100 square metres in terrain that has an 11,000-metre level
of vertical relief. Just achieving the pinpoint touchdown capability would help those probes that would follow. As well as
crater Shackleton it could also head for the six-kilometre-
high Malapert Mountain, 120 kilometres from the pole, from
which the Earth can virtually always be seen, thus allowing
continuous communications for any future outpost in the
region. The regions of almost perpetual light are in direct
view of Malapert.

The project seemed to be gaining momentum in March
1998 as representatives of ESA's fourteen member nations
met to consider the future of EUROMOON 2000 just weeks
after the announcement that Lunar Prospector had found ice.
Before the meeting the project was said to have the support
of Rodotà and Britain's science minister. But after two days of
discussions, with most of the money for EUROMOON 2000

secured from private sources, ESA's council turned down the funding request for £32 million. After the meeting Ockels said, 'We are basically sitting here crying. I don't know what happened. When the director general of ESA asked who supported EUROMOON 2000 there was silence. Nobody said anything. Everybody got scared. I was sitting there thinking: I can motivate industry. I can motivate the public. But the politicians just don't get the point.' Ockels died in 2014 having not seen his dream fulfilled.

If ESA had gone for the project, it would have put them at least a decade ahead of anyone else and enabled them to lead on such a project. Some believe that if there really is ice there the Moon could become the solar system's filling station and be at the heart of manned exploration of the solar system. According to NASA's Ben Bussey, 'The poles are an enabling factor for future exploration, both robotic and human.' Space agencies and private companies are now turning to the Moon, but ESA had the chance to lead the way 20 years ago. The lack of interest in EUROMOON stalled lunar exploration for a few years but soon the need to go back became irresistible. A new wave of lunar probes would be dispatched in the first decade of the new century.

THE LIVING DAYLIGHTS

---•---

In the early part of the 2000s the European Space Agency launched SMART-1 (Small Missions for Advanced Research in Technology) as part of a program to build small and relatively inexpensive spacecraft. It was propelled by a novel thruster which used xenon as a propellant. It used an electrostatic field to ionise the xenon and accelerate the ions to high speed. The power for the electrostatic field came from solar panels. It was not very high thrust, but it was constant, allowing speed to be accumulated. Even so it took fourteen months to reach the Moon. Over the following twenty months it mapped the Moon's minerals. On 3 September 2006, having reached the end of its mission, it was deliberately crashed on the Moon with the impact flash visible, by telescope, from Earth.

In October 2007, Japan's Kaguya was the largest lunar mission since Apollo times to orbit the Moon. Three satellites, an orbiter, relay spacecraft and a radio satellite lasted for over a year and took the first high-definition imagery of the lunar surface. What I most remember about this mission is the series of spectacular images it took of the full Earth hovering above the lunar south pole: the blue-and-white beauty of the Earth contrasted with the Moon's monochrome shadows and illuminated crater ridges. The stills are amazing and the movie from which they come is more so. Here was the land we were heading towards.

Kaguya produced a global topography map and the first detailed gravity map of the far side, as well as the first optical observation of the permanently shadowed interior of the crater Shackleton. Using high-sensitivity cameras Kaguya was able to use what very little light there was inside the crater. To the human eye it was all dark but to Kaguya's eyes the steep sides could be seen and jumbled, hilly terrain at its floor.

Just over a year later India's first Moon probe, Chandrayaan-1, entered lunar orbit and used its Moon Mineralogy Mapper to scan the surface, confirming the presence of iron. In the Orientale basin there were indications of iron-bearing minerals such as pyroxene. The X-ray signatures of aluminium, magnesium and silicon were detected after a solar flare caused them to fluoresce. Unfortunately, Chandrayaan's planned two-year mission was cut short in August 2009, but during its time it released a Moon Impact Probe into Shackleton crater. The location of the impact was called Jawahar Point and the ejecta thrown up confirmed the presence of ice. The importance of Shackleton crater was growing.

In one set of observations Chandrayaan imaged a lunar rille near the equator, formed by the collapse of an ancient lava flow. Looking more closely, an uncollapsed segment was found that formed a large cave measuring about two kilometres in length and 360 metres in width. It could be used as a base, offering good protection against solar flare radiation. Looking back at Kaguya's data, scientists found many more caves.

It was the US spacecraft the Lunar Reconnaissance Orbiter (LRO) that revolutionised what we knew about the Moon. Reaching orbit in June 2009, its main task was to search for landing sites. The data obtained by LRO has been described as essential for planning NASA's future human and robotic missions, identifying safe sites and locating potential resources. It produced a very impressive database, but Shackleton remained the prime candidate for study. Additionally, LRO's image of Earthrise over Compton crater has become a classic of science imagery.

Then came the LCROSS (Lunar Crater Observation and Sensing Satellite) mission that crashed two probes – a mothership and an empty rocket upper stage – into the permanently dark depths of Cabeus crater, about 100 kilometres from the south pole. The mothership was the second of the two to crash and it detected water vapour in the debris cloud thrown up by the first. In the plume grains of pure water ice were detected.

Astronomers were focusing on Shackleton crater. Among them was Professor Maria Zuber of MIT's Department of Earth, Atmospheric and Planetary Sciences. 'We decided we would study the living daylights out of this crater,' she says. 'We were able to make an extremely detailed topographic map.' It took 5 million pulses from the Lunar Reconnaissance Orbiter's laser ranging device during more than 5,000 flyovers to produce a detailed topographic map of Shackleton, far superior to the fuzzy image obtained by Kaguya. It revealed that the sides of the crater are smooth and steep, exceeding 35° in places. Zuber describes the crater's floor as 'extremely rugged. It would not be easy to crawl around in there.' The temperature never exceeds 100° above absolute zero and the data suggests up to 22 per cent ice in the top 1mm of regolith.

As we have seen, what particularly interests scientists about Shackleton is not just its dark interior, but also the sunlight shining on its rim, and the LRO data was used in the search for a 'Peak of Eternal Light'. Alas, no such place exists. In a complementary study Ben Bussey has used Kaguya data and also found no areas of constant illumination, but he did find two areas that were particularly interesting. One was on the rim of Shackleton and the other was on a ridge 15 kilometres away – illuminated for more than 90 and 98 per cent of the time respectively.

We have studies of sunlight and ice but we have reached the limit of what we can find out about Shackleton using the satellites we have. The next step must be to land a probe on its rim. Ben Bussey says that to really settle the question of lunar ice, a lunar lander will be required: 'We need to learn exactly what is the nature of the polar volatiles, and how difficult they are to extract. This information could be acquired with a rover.'

Despite the fact we have done it many times in decades past, landing on the Moon is not something that can be taken for granted. On 11 April 2019 the Israeli-built Beresheet (the name means 'In the beginning') was moments away from being the first mission to land on the Moon with a lander built by a private company, as well as the first Israeli mission to land on the Moon. It was firing its thrusters in a braking manoeuvre as it headed to the northern regions of Mare Serenitatis when a gyroscope malfunctioned and there was no time for ground controllers to do any resets because of a sudden loss of contact with the probe. By the time contact was regained the thrusters had been out of action for too long. Desperately they were fired but Beresheet could not be slowed down enough and smashed into the surface. Soon afterwards the LRO overflew the crash site and found a new small crater. The Israelis vowed they would be back.

Nobody had landed at the south pole – at Shackleton or anywhere near it; it would be a big prize for any nation or company that did it. India was determined to be first.

Chandrayaan-2, India's second lunar exploration mission, reached the Moon on 20 August 2019 and prepared to release the Vikram lander, carrying the Pragyan rover, towards the south polar region, aiming for a high plain between the craters Manzinus C and Simpelius N. All was going well until the craft reached an altitude of 2.1 kilometres, when it veered off-course and crashed. Investigators suspected a software glitch. The debris was seen spread over a wide area. Soon after, Prime Minister Narendra Modi, who was present to watch the historic moment, addressed scientists and said they shouldn't be disheartened: 'Nation is proud of you, be courageous. You have given your best and have always made India proud.'

There is one country that I haven't yet mentioned that has a very ambitious Moon program and will undoubtedly play a major part in its future exploration: China. It has already sent Chang'e 1 and 2 to lunar orbit in 2007 and 2010 respectively. In December 2013 Chang'e 3 soft-landed on the Moon, becoming only the third nation after the USSR and USA to do so. It deployed a rover called Yutu (Jade Rabbit).

Chang'e 4 was an even more dramatic mission – the first to the far side of the Moon. To get the probe to a region of the Moon that is never in sight of the Earth required a masterclass of flying in cislunar space. To communicate with Earth the Chinese team placed a communication relay satellite, called Queqiao (Magpie Bridge), into an orbit around the Moon's so-called Lagrange 2 gravitational balance point. In this position 65,000 kilometres beyond the Moon the satellite can take advantage of being in a place where the gravity of the Earth and the Moon balance, allowing the satellite to retain a stable position in space. It was the first lunar relay satellite at this destination. China also launched two microsatellites to lunar orbit: only one succeeded, functioning for a while before being ordered to crash on the Moon.

Shortly after sunrise over the Von Kármán crater Chang'e 4 landed on 3 January 2019. China released video of the final stages of the landing as Chang'e 4 self-guided to the surface, avoiding rough terrain. The landing site is of scientific interest. Von Kármán crater is in the South Pole–Aitken basin, one of the most significant large-scale structures on the Moon, its deepest and oldest impact scar. You could place Mount Everest on its floor and its summit would not even come close to poking out of it. The impact gouged material from the Moon's mantle and brought it to the surface. Afterwards the lava seeped in from below and flooded the crater.

Twelve hours after landing it lowered a ramp and deployed the small six-wheeled Yutu-2 rover, carrying a camera, spectrometers and ground-penetrating radar. After a few days Yutu-2 went into hibernation, for it could not operate in the long lunar night.

The lander contained an experiment that got a lot of publicity. It was its Lunar Micro Ecosystem which contained seeds and insect eggs to see if they could respectively sprout and hatch on the Moon. It housed six types of organisms: rapeseed, potato, cottonseed, a flowering plant, yeast and fruit fly eggs. Within a few hours of landing, the temperature inside the experiment was adjusted to 24°C and the seeds automatically watered. China said that the potatoes, rape and cottonseed had sprouted, but provided pictures of only the cottonseed. It seems the experiment ended prematurely due to a heater failure. During its first lunar day Yutu-2 travelled 120 metres. It is still operating, having broken the record held by the USSR's Lunokhod 1 in December 2019, and has made some fascinating observations. In June 2020 both the lander and the rover were still going strong having just started their fifteenth lunar day.

For many years China has been seen as a dark horse in space exploration with its slow and steady accumulation of capabilities. A few years ago, Michael Griffin, when he was head of NASA, caused a few to feel uncomfortable when he suggested that China would land humans on the Moon before the United States returned. 'I think when that happens, Americans will not like it, but they will just have to not like it.' For a while it seemed as though he was right, but not any more.

China's long-term goals in space are clear, a human mission to Mars, but first they want a base on the Moon by the mid-2030s. Although in the past few years they have said that they will land on Shackleton crater, I believe the decision will not be as simple when seen from the perspective of the 2030s. Despite the attractions of Shackleton, with the Americans, Russians, European Space Agency, India and other countries also being interested in it, China might want to establish a base elsewhere – and there is really only one other place they can go: to the permanent shadows at the other end of the Moon. The north pole is on the northern rim of Peary crater and data indicates that the small craters on its southern floor, as well as nearby Hermite A and Hinshelwood, all contain ice. Unfortunately, none of these sites has a clear line of sight to the Earth for communications, but that could be remedied with a relay dish.

There is another place that might interest them. Pits have been seen on the north-eastern floor of Philolaus crater, a crater 70 kilometres in diameter, about 550 kilometres from the pole. Philolaus crater is relatively young, one of the rare large craters formed within the last 1.1 billion years or so. The pits are small rimless depressions, typically 15–30 metres with completely dark interiors. They are thought to be collapsed, or partially collapsed, lava tubes, underground tunnels that were once streams of flowing lava.

Some want to go to the Moon very soon, and in numbers.

Elon Musk is a giant in the space business and the company he has developed, SpaceX, is undoubtedly a magnificent company with achievements that have changed the course of space history. It built Falcon 1 in 2008, the first privately funded liquid-propellant rocket to reach orbit. It was the first private company to successfully launch, orbit, and recover a spacecraft, which it did with its Dragon spacecraft in 2012. It became the first private company to send a spacecraft to the International Space Station, the first to reuse an orbital rocket (Falcon 9 in 2017), and the first private company to launch an object into orbit around the Sun (Falcon Heavy's payload of a Tesla Roadster in 2018), and the first private company to launch astronauts into orbit. SpaceX has become a thriving commercial operator in putting satellites into space, undercutting its rivals in price, and it has done it all in just a few years. It has made NASA and everyone else look slow. It has shown that things can be done swiftly and well.

After demonstrating that the Dragon capsule could resupply the space station, thoughts inevitably turned towards upgrading it for humans. NASA awarded SpaceX a development contract in 2011 to do this, to transport astronauts to the ISS and return them safely to Earth. That was the year the space shuttle was retired, leaving America with no human access to space of its own, having to rely on Russia's

Soyuz capsules. SpaceX conducted the maiden launch of its Dragon 2 spacecraft on a NASA-required demonstration flight on 2 March 2019.

The first crewed Dragon capsule, named Endeavour, was launched on 30 May 2020 commanded by Doug Hurley, who had commanded the last flight of the space shuttle in 2011. His co-pilot was Robert Behnken who had also flown two space shuttle missions. Endeavour docked with the International Space Station some 18 hours later while over the northern border of China and Mongolia.

When the space shuttle retired NASA had to pay Russia up to $80 million a flight to take its astronauts to the space station. This was politically unacceptable as, despite the friendly relations enjoyed by US and Russian space officials, it was realised that President Putin could cut off the service at any time. NASA primed SpaceX with over $3 billion to develop the Dragon capsule and its operation and following its successful first crewed mission astronauts will be regularly ferried to and from the space station, including the actor Tom Cruise who is working with NASA to make a movie there.

NASA did not just rely on SpaceX, they also backed the Boeing company with its Starliner capsule, which is slightly larger than the Apollo Command Module and smaller than the Orion capsule. It's designed to be reused up to ten times and can be launched on many rockets currently in use. It is lagging behind SpaceX's Dragon as during its first uncrewed orbital test on 20 December 2019 it suffered a malfunction when its onboard computer clock was reset in orbit, making the capsule's computer think it was at the terminal phase of the mission instead of at its beginning. Despite this, the capsule managed to land prematurely.

Subsequently, however, it was reported that there had been two further software errors. One of them would have prevented docking with the space station. Each of the errors could have resulted in the destruction of the capsule had they not been caught in time. Boeing has since announced there will be another uncrewed orbital test in October or November 2020 before astronauts will ride on it to the space station in 2021.

But Elon Musk wants to go further than the space station. In 2019 he told *Time* magazine that he wanted to land on the Moon within two years and he's building a spaceship to do it. He calls it the Starship. He says it will fly to the Moon, to Mars and beyond. Standing in front of the 50-metre-tall 200-tonne prototype, he said it's 'the most inspiring thing I've ever seen'.

He already has a customer for a trip around the Moon. Yusaku Maezawa is a Japanese clothing billionaire who in 2018 started the #dearMoon project to take himself and a handful of artists to the Moon and back using the Starship. The project's literature says it could happen in 2023, before the landing of Artemis 3. Six to eight artists would travel with Maezawa on a six-day mission in the hope that new art would be created as each interprets the mission in their own way. The art would then go on a tour to promote world peace.

It's not going to happen, and here's why. The Starship is in its infancy and there are major technological hurdles to be overcome in getting it into space, let alone into orbit.

In July 2019 it made a test hop of 20 metres; a month later it made a second hop and landed 100 metres away from the launchpad. There is a long way to go. Starship is a new kind of vehicle. It's not just a capsule with some thrusters, as has been the case up to now; it's the upper stage of a rocket with an integrated capsule. There will be no abort procedure with it like there is for other craft, whereby the capsule can separate swiftly from the explosive rocket. Launching Starship as the upper part of a two-stage rocket will require many, many tests, especially of the re-entry and landing, which is the most severe test of any spacecraft. There will have to be the integration of flight controls, training of the pilots, development of flight manuals and procedures. Then there are the development and integration of life-support systems, training and medical tests on the passengers, insurance and legal matters. After that there would have to be an uncrewed flight around the Moon. All this in two years! As I write this, no artist has said they're going. Given the overwhelming problems, tight schedule and risk, you are not going to get any rock stars signing up.

Musk sees the Starship as the beginning of something big, what he calls an Interplanetary Transport System. If it were refuelled in Earth orbit it could have the ability to reach Mars, and in one design it has a 1,000-cubic-metre pressurised volume that includes 40 cabins, a galley, a lounge and a solar flare shelter. But as I have said, designs are one thing, turning them into reality is another. This Starship will not get us to Mars, as we shall see.

BUILDING
THE MOON BASE

———•———

The crew of Artemis 3 will land at a site that has already been visited. Scattered around them will be many spacecraft sent to gather data and to put in place equipment they will need. For example, the farthest the astronauts will be expected to walk in their new moon suits will be about half a mile, so they will need a rover to extend their range, just like the astronauts of Apollos 15, 16 and 17. It will have been waiting for many months, having already undertaken several forays across the rim of Shackleton, guided by controllers on Earth. At the time of Artemis 3's landing the rover's camera will be looking for its approach, seeing it first as a bright point of light coming over the crater and following its landing nearby in real time for the audience on Earth. In addition, sixteen experiments will be placed on Shackleton's rim in 2021–22 by two landers. One is being developed by Astrobotic of Pittsburgh, which will send eleven experiments on a lander launched by a Vulcan rocket built by the

US United Launch Alliance. The Peregrine lander will study the chemistry of the surface and the radiation environment, even injecting contaminants into the regolith to see what happens. Intuitive Machines of Houston are building the other lander, which will be launched on a SpaceX Falcon 9 rocket and will test communications and use stereo cameras to look at the debris kicked up by its landing plume.

With such a tight deadline the construction of the crewed lander is also underway. Many companies have designs they have submitted to NASA, who chose three of them for further development. SpaceX's Starship is by far the largest and most complicated of the three choices and it seems unlikely that it could be ready on the timescales required but its inclusion is for what it might be able to offer in the future. Also chosen was a low-slung lunar lander designed by Dynetics and a three-stage lander designed by the so-called National Team that is led by Blue Origin.

It's possible that all three designs will land on the Moon. The first company to complete its lander will carry astronauts to the surface in 2024, and the second company will land in 2025. Typically, spaceflight hardware can take six to eight years to develop so the development of the landers will be done in an unprecedently short time. The lander, by any design, will be larger than the lunar module that took Armstrong and Aldrin onto the Moon.

Since time is of the essence to get the task done, NASA has been relaxing its requirements for the lander, even if only for a while. In addition to the original requirement that the landers could touch down anywhere on the Moon, NASA originally required them to be refuellable. Many engineers were worried by this, so NASA agreed to remove it so that industry had greater flexibility in their designs. 'They were absolutely right,' said Lisa Watson-Morgan, the Human Landing System program manager at NASA's Marshall Space Flight Center in Huntsville, Alabama. 'We are operating on a timeline that requires us to be flexible to encourage innovation and alternate approaches. We still welcome the option to refuel the landing system, but we removed it as a requirement.'

The crew of Artemis 3 will find support spacecraft positioned around their landing site. They will also benefit from a wide range of craft that will be sent to the Moon during the run-up to their mission. There are plans for an ambitious vehicle called VIPER (Volatiles Investigating Polar Exploration Rover) to be delivered to the surface of the Moon as early as December 2022. It will be the first mission able to move around and investigate the true nature of the ice deposits and allow plans to be drawn to turn such deposits into a resource available to astronauts. The precise area around the south pole that VIPER will explore has yet to be decided. The vehicle will be powered by solar panels but once it drives itself into a permanently shadowed location, it will operate on battery power alone and will not be able to recharge them until it returns to a sunlit area. Its total operation time will be approximately 100 Earth days. The scientific return from VIPER would certainly be interesting, but it seems that industry is nervous about the project, chiefly because it could not at the moment be set down on the lunar surface as accurately as NASA would like. If, for example, it were to be sent to the lip of Shackleton, a 100-metre landing error due to whatever cause could see it tumbling into the depths of the crater.

At around the same time, India will try a landing once again, as will the Japanese with their first mission to the lunar surface. It's called SLIM (Smart Lander for Investigating the Moon) and it will be going to a particularly interesting place that has an analogue on Earth.

When in Iceland, a journey into Raufarhólshellir is a unique experience and a great opportunity to explore the inner workings of a volcanic eruption, as one walks the path taken by an eruption over 5,000 years ago. The total length of the tunnel system is an impressive 1,360 metres, the main tunnel being 900 metres long. It is up to 30 metres wide with headroom up to 10 metres high, making it one of the most expansive lava tunnels in Iceland. At the end of Raufarhólshellir the tunnel branches into three smaller tunnels where magnificent lava falls are seen. Raufarhólshellir is a lava tube: a conduit formed when an active low-viscosity lava flow develops a continuous and hard crust, which thickens and forms a roof above the still-flowing lava stream. SLIM is heading to a similar but far more ancient place on the Moon. It's called the Marius Hills Hole and it leads to the question of whether humans could live beneath the surface of the Moon. It was in 2009 that Japan's Kaguya spacecraft observed a curious hole beneath the Marius Hills region, possibly a skylight to an underground lava tube, carved out by lava billions of years ago and left empty as the lava drained away. Follow-up observations by NASA's Lunar Reconnaissance Orbiter (LRO) indicated that the Marius Hills Hole extends down nearly 100 metres and is several hundred metres wide. More recently, ground-penetrating radar data from Kaguya has been looked at again to reveal a series of intriguing secondary echoes that suggests that more extensive lava tubes exist under the Marius Hills and might extend down kilometres below the surface, making them large enough to house cities. Such tubes could provide shelter for a future Moon base.

With two orbiters and two landers under its belt, China is wasting no time in its plan to explore and colonise the Moon. At the end of 2020 Chang'e 5 will hopefully bring a piece of the Moon back to Earth. It will be China's first sample return mission, aiming to return at least 2 kilograms of regolith and rock samples. The planned landing zone is Mons Rümker in Oceanus Procellarum, located in the north-west region of the near side of the Moon. Rümker is a fascinating place to set down. It is the largest contiguous volcanic edifice on the Moon, composed of a series of overlapping lava flows. There are a lot of lunar domes in the vicinity, formed by erupting and cooling lava. It has a strong spectroscopic signature of basaltic lunar mare material. The mission will have four elements: the lander will collect about 2 kilograms of samples from 2 metres below the surface. These will be placed in an attached ascent vehicle that will be launched into lunar orbit. The ascent vehicle will make an automatic rendezvous and docking with an orbiter that will transfer the samples into another capsule for delivery to Earth.

After a long time out of the business, the Russian Federal Space Agency – Roscosmos – are also planning a lunar-landing mission: Luna 25 in 2021. It will land near the lunar south pole at Boguslavsky crater. Currently it is planned as the first of several missions, culminating in Luna 27, a joint European Space Agency lander that will touch down in the same region as Artemis 3 at about the same time. Before that, Luna 26 will go into a polar orbit to relay communications between Earth and future landers.

The human landing is the one that everyone is waiting for; it's only been done six times. Touchdown on the Moon, or indeed on any other world, will always be a tense time. The landing of Artemis 3 will be far different from those of the Apollo missions. Only crude surface information was available for them. Artemis 3 will have a detailed map of the touchdown zone on the rim of Shackleton, as well as auto-targeting based on the relative positions of the equipment pre-positioned there.

When Armstrong set foot upon the Moon, he said the words that almost everyone knows. He wasn't prompted or instructed about what to say but thought about them on the way there. Somehow, I don't think it will be the same when the 13th person steps onto the Moon. The two astronauts will remain on the surface for six-and-a-half days, double the time of the longest Apollo mission. They will carry out up to four moonwalks, some using the rover, travelling up to 15 kilometres from the lander. They will carry out a variety of scientific observations, the most important being a geologic survey of the area and determining what the regolith, that is, the surface dust and rock, is like. They will peer into the dark depth of Shackleton and shine a torch into it, but they will stay clear of the edge. They will look around the lander for small craters of about a metre in size that could have small regions of permanent shadow. They will gather samples and carry out analysis of them.

But with regard to bringing moon rocks and dust back, there is at present a slightly embarrassing problem. One of the limitations on returning samples is the Orion spacecraft itself, as it does not have any designated space for rocks. At the moment no one knows how much it will be possible to return to Earth. Initial drafts of NASA's Human Landing System procurement – the plan that involved using the Gateway on the first landing mission – called for the return of 100 kilograms of samples, including the sample containers. The final version reduced that requirement to 35 kilograms: 26 kilograms of samples and nine kilograms for the containers. Before the plan was changed it seemed that more rocks could be delivered to the Gateway than could be delivered to Earth. Not using the Gateway makes the problem worse as only the amount of rocks that can be accommodated by the Orion capsule will be brought from the lunar surface. Transporting rocks in the Orion capsule would obviously be the best thing to do but there may be the need to study alternative ways, such as a robotic sample return vehicle delivered to the surface of the Moon. It would not be an ideal solution. It would not be ideal for the samples brought back to Earth from Artemis 3 to be only slightly more than the 22 kilograms returned by Apollo 11 with its single two-hour moonwalk.

Apollo 17 returned about 110 kilograms collected during three lunar excursions.

After Artemis 3 has put people back on the Moon, it is hoped that there will be several more missions, at roughly yearly intervals at first. Artemis 4 is scheduled to follow in 2025, followed by the delivery of a pressurised rover as early as 2026 according to NASA, but more realistically a few years later.

For the first few Artemis missions the crew will live inside the lander, but soon the time will come to build on the Moon. For a long time, it was thought that a Moon base would be made out of preformed modules, initially of cylindrical shape, made of metal and composites and outfitted with the required equipment and life-support systems. They would be landed on the surface and astronauts would connect the modules to an airlock and power source and link them with flexible tunnels. More inflatable domes could be added to the outside to provide extra lab space. Subsequent missions would bring more modules and possibly a tractor to pile regolith over the outside, providing much-needed radiation protection, which we will consider in the next chapter.

For most of the exploration of the Moon, bases were designed to be positioned near to the equator, as it's easier to land there and the Sun is high in the sky for much of the two-week-long lunar day, allowing solar cells to provide the required power; but the two weeks of night-time present a severe problem. With no atmosphere, the temperature is going to get extremely cold very swiftly and stay that way. The Moon base needs energy when the solar panels are in darkness. The solar cells could charge up batteries as they do on the International Space Station, but battery technology is not as advanced as you would imagine. They are very bulky, need a lot of maintenance, are prone to breaking down and will eventually need replacing. A small nuclear reactor has often been seen as part of the solution but there are the obvious problems of launching it from Earth into orbit, from where it has to be transported to the Moon. Surviving the fourteen-day lunar nights became a fundamental factor in the design of a base near the equator, but then came polar thinking.

As we have seen, the temperature ranges in the polar regions are far less dramatic than at the equator, the Sun can be available for long periods, and there is also ice. Solar power is a far better option at the poles. There has also been another change to our plans for the Moon base. Why ship habitat modules to the Moon at all? After all, the Moon has all the materials needed to build shelters. Why not use them? There is a way.

The International Space Station has been home to astronauts continuously for more than nineteen years and it's expected to be occupied for perhaps another six years or more. Astronauts live and conduct scientific research using dozens of special facilities: there are places to sleep, eat, relax, exercise and work. To make this possible, more than 3,000 kilograms of spare parts has to be sent to the station every year.

Such a supply logic works well for the space station, orbiting 400 kilometres above Earth with regular cargo missions, but it won't work for the Moon. This is where 3D printing technology could provide a solution. The commander for the International Space Station's Expedition 42, Barry Wilmore, installed a 3D printer in 2014 and carried out a calibration test. It worked well. Soon afterwards, ground controllers sent the printer the command to make the first printed part: a faceplate of the extruder's casing. It worked. The printer fed a continuous thread of plastic through a heated extruder and onto a tray layer by layer to create a three-dimensional object. 'This first 3D print is the initial step toward providing an on-demand machine shop capability away from Earth,' said Niki Werkheiser, the project manager. 'The space station is the only laboratory where we can fully test this technology in space.' The printer also made an antenna part, and an adaptor to hold a probe in an air outlet on the station's oxygen generation system.

Already, Japanese companies can assemble a house using robots. Everything required is positioned next to a robot which has arms running on tracks adjacent to the site to be built upon. First the robot digs the foundations, fills them with concrete and then puts in place the already built walls, doors and floors, which contain all the required fittings and connections. The same thing could happen on the Moon.

On the lunar surface the basic building material will be the regolith, which will be collected and processed, possibly by adding sulphur or magnesium oxide, so it has the consistency of sludge. It's a technique that has the advantage of not requiring water. The robot will pump this material into one of its arms and add it onto a prepared platform, building up a cylindrical or dome structure around an inflated inner liner already outfitted with communal areas, laboratories, bunk areas and an airlock. Once the external walls are complete, regolith would be added to make the wall thickness over 150 centimetres as required for radiation shielding.

One of the problems the Moon base will face after very few landings is that of debris. Close examination of the landing and take-off videos of the Apollo missions show that the Apollo lunar modules dispersed at least a ton and possibly several tons of dust, sand, and rocks with each landing. Rocks of 4–10 centimetres in diameter were seen blown by the lander's exhaust, travelling as fast as 11–30 metres per second as they passed through the field of view of the lander's camera. Astronauts even saw some of this dust passing over the horizon. Calculations suggest silt-size particles travel at 1–3 kilometres per second. Since the Moon's escape velocity is 2.38 kilometres per second, some of these particles travelled all around the Moon and some even escaped to go into orbit around the Sun. Clearly any spacecraft, rover or any equipment nearby could be damaged by each and every landing and take-off.

There are several ways to mitigate these effects. The landing zone could be placed behind a small undulation, or barriers could be constructed, although they would need to be sited carefully as, owing to the Moon's lack of atmosphere, particles would just bounce off them and could cause problems elsewhere, possibly even striking and damaging hardware from above as well as from the side. Perhaps the best, longer-term solution is to modify the landing site itself using a process called sintering. It's the process of combining materials using heat or pressure and is different from melting in that the materials do not combine after liquifying. If the surface is loose dust, one can imagine a modified lunar rover fitted with a vibrating or compacting arm to treat the surface. There are other ways to sinter the regolith, using lasers, microwaves or infra-red heating. A rover could even act as a bulldozer, scraping the surface flat. Perhaps the surface could be sprayed with a polymer to bind the surface – NASA has studied such a technique – though a drawback is that the polymer would have to be transported from Earth.

But when the base is beginning to be established and we learn how to live on the Moon just as we have learned to live on the International Space Station, we will turn our thoughts to the science possible on the Moon. Consider how it differs from the Earth: low gravity, no magnetic field, no atmosphere (a good-sized room holds as much gas as the entire lunar atmosphere), high vacuum, low or high temperature depending upon whether you're in shadow, seismic stability, no radio interference on its far side, and total sterility.

The Apollo 16 astronauts set up an ultraviolet scope in 1972. It looked at several UV sources including the Earth's upper atmosphere and nearby galaxies. As a follow-up NASA scientists designed LUTE, the Lunar Ultraviolet Telescope Experiment. It would have been placed on the Moon by an unmanned lander, but Congress never funded it. With its 1-metre mirror, LUTE would not have had the ability to move but would stare at a swath of sky 1.5° wide every 28 days as the Moon rotated.

Orbiting observatories may be cheaper than lunar ones and, in many respects, have advantages, but the Moon has some important things to offer astronomers looking for a place to put their instruments. It is very stable for arrays of telescopes that need to know their relative positions to minute precision. Optical telescopes on the lunar surface, separated by just 1 kilometre but linked, could have a resolving power hundreds of times better than the Hubble Space Telescope. They could see the width of a coin at 2 million kilometres. One group of scientists has proposed the Lunar Synthesis Array: an interesting concept involving two concentric rings of 1.5-metre telescopes. The outer ring would consist of 33 telescopes in a circle 10 kilometres across. The inner ring would be 0.5 kilometres across and would contain only nine instruments. It would have a resolving power 10,000 better than the Hubble Space Telescope and could detect Earth-like planets orbiting other stars.

The Moon's far side could be the quietest place in the solar system for radio astronomy, allowing us to study very low frequency signals, below 15 megahertz (wavelengths longer than 20 metres). These frequencies never get through the Earth's atmosphere and so are one of the few remaining unexplored windows in the electromagnetic spectrum. In the Moon's one-sixth of Earth's gravity, telescopes could be larger with lighter mirrors and if positioned at the poles could carry out longer exposures of certain celestial objects not possible on instruments elsewhere.

There could also be an underground isolation facility. Situated in a man-made cavern ten metres below the surface,

it would be shielded from all radiation except neutrinos and the radioactive decay products from naturally occurring potassium, uranium and thorium.

What could we put in such an isolation facility? We could have a back-up global seed vault like the one at Svalbard on the Arctic island of Spitzbergen. Its aim is to preserve a wide variety of plant seeds to ensure against their loss in other gene banks during large-scale regional or global crises. We could store a library of genome sequences and proteins. We could add histories, commentaries, holograms of art, volumes of plays, all catalogued and demonstrated by an AI guide.

It is not infeasible that humanity could be destroyed by a giant asteroid impact or a militant virus. Or if not destroyed, then sent back tens of thousands of years into pre-civilisation days. If this happened and we clawed our way back, perhaps rumours of a library on the Moon would have survived, wrapped in and embellished in mythology by humans thrust into a pre-industrial age for tens of thousands of years.

Perhaps we will build such a library on the Moon in the next 50 years. If we do, its AI curator should be named after one of the greatest libraries we ever built and lost on Earth. After the catastrophe it would be sought out by the next wave of humans to walk upon the Moon, and if they found it they would be greeted by its guardian and interlocutor, Alexandria.

THE LIFEGUARDS
OF SPACE

———•———

Danger is ever-present in space. At NASA's Johnson Space Center in Houston, and in Colorado where the Space Environment Center of the National Atmospheric and Oceanic Administration is situated, scientists monitor space conditions for the astronauts on the space station. These space environment officers have been called the lifeguards of space. Every day they look at the conditions in space and produce a forecast addressing what is happening on the Sun: whether harmful charged particles are on their way to the Earth, and if so when will they arrive at the outer guard of space-monitoring satellites beyond the Moon. On Earth we are protected from solar particles by our planet's magnetic shield – the magnetosphere. The exposure of the crew of the space station depends upon a number of factors such as the structure of the station, what it's made of, the altitude, the inclination of its orbit, the status of outer zone electron belts that encompass the Earth, the interplanetary proton

flux, geomagnetic field conditions, solar cycle position, and, if applicable, the start time and duration of any spacewalk. If the energetic particle flux from the Sun, or 'space weather', shows signs of increasing, the scientists could advise the crew to postpone any spacewalk.

Space radiation is a serious problem for space travellers. During the Apollo missions we were lucky not to lose a crew because of it. On 7 August 1972, between the Apollo 16 and 17 missions, a huge solar flare was detected by the Big Bear Solar Observatory in California. This particular flare – known as the seahorse flare for the shape of the bright regions – initiated a strong event. Had a crew been in space they would have experienced high levels of radiation that could have been fatal.

Radiation readings on the lunar surface carried out during the Apollo missions were limited, covering only a small energy band and not sufficient to calculate the human exposure. More recent work performed by the Radiation Dose Monitoring instrument on the Chandrayaan Moon mission and the Cosmic Ray Telescope on the Lunar Reconnaissance Orbiter have yielded better results, but they are still orbital measurements. We need data from the surface.

As well as the flares, sometimes the Sun throws off clouds of material – called coronal mass ejections – containing a billion tons of material. It's thought they play a major role in the Sun's most powerful radiation events, called solar energetic particles or SEPs. They are almost all protons, flung at such high speeds that some reach Earth, 150 million kilometres away, in less than an hour. It's rather like when a

high-speed boat goes through water, you can see the wave ahead of it. The shock waves ahead of fast coronal mass ejections accelerate particles before them. SEPs are dangerous because they pass straight through skin, releasing energy and damaging cells and DNA, increasing the risk of cancer.

The International Space Station in low-Earth orbit is generally safe because it orbits within the Earth's magnetic protection, and the hull helps shield the crew. But beyond the Earth's magnetic influence lies danger. There are short-term and long-term ways to deal with it.

Planned spacecraft will not have built-in massive dedicated radiation shielding since it would make the craft too heavy to launch. Every item in the spacecraft will therefore have an additional purpose to contribute to radiation protection if possible. If during a future Artemis mission a solar radiation event were to occur while the astronauts were in the vicinity of the Moon or in transit, mission control would tell them to build a shelter, surrounding themselves with as much mass as possible — whatever is available. They would need to redistribute it to fill in areas that are thinly shielded and the crew would be directed towards more heavily shielded areas. The more mass between the crew and radiation, the more likely that dangerous particles will deposit their energy before reaching them. Engineers believe that if the Sun erupted with another storm as strong as the Apollo era's, the Orion crew would be safe.

As the coastguards of space know, protecting astronauts from solar storms requires knowing when such a storm will occur — but they are difficult to predict. We do not know enough. Even seeing an angry-looking active region of the Sun's surface isn't enough. Only a small fraction of them result in particles that are hazardous to astronauts. Even if they do, it's hard to predict where they will go. Magnetic fields emerging from the Sun guide their direction into space, but those fields are convoluted and unpredictable.

There are also seasons for flares. The Sun goes through an eleven-year cycle of high and low activity. During solar maximum, there are many sunspots: regions of high magnetic tension that can become unstable, explosively converting magnetic energy into heat, causing a flare. During solar minimum, when there are few to no sunspots, eruptions are rare, though the Sun can always throw up surprises. Astronomers who monitor the coming and going of sunspots note that the Sun has just gone through an unusually long minimum and estimates are for a relatively weak forthcoming cycle ending in the early 2030s. This prolonged solar minimum and its lower incidence of flares might be good news for the first wave of lunar explorers, or possibly not.

There is another kind of space radiation that is even more dangerous. Galactic cosmic rays are high-energy particles travelling at near-light speed from exploded stars elsewhere in the galaxy and are more powerful than the most energetic solar particles. They are comprised of elements like helium, oxygen or iron and are able to break apart atoms when they collide with, say, the metal walls of a spacecraft or the DNA of an astronaut. The impact sets off a shower of more particles, adding to the health problems. A shield that would protect a crew from SEPs would not protect from cosmic rays.

This type of radiation is a major problem for long-duration missions like the journey to Mars. On Earth, the contribution to the annual terrestrial radiation dose of 2.4 milli-Sieverts (mSv) – a measure of radiation dosage – by cosmic radiation is about 15 per cent. It's much worse on the surface of the Moon, being roughly 380 mSv at solar minimum and 110 mSv at solar maximum, much higher than if a person spent six months onboard the space station, which would result in 80 mSv.

It's thought that the very worst case of radiation an astronaut above the magnetosphere would be exposed to is 1 Sv and NASA has stipulated that is the maximum dosage an astronaut is allowed during their career. The only efficient measure to reduce radiation exposure is the provision of radiation shelters. The cosmic ray flux is also related to the solar cycle. At solar minimum the rays easily penetrate the Sun's magnetic field. But during solar maximum, the Sun's magnetic bubble strengthens, protecting us and the other planets from them.

DIGGING, DRILLING AND DRIVING ON THE MOON

———•———

As we reach the 2030s the US will have much of the equipment it needs on the Moon, as well as the phase one architecture of the Moon base and the transport and communications infrastructure for a sustained presence, but the extraction of water from the crater's ice will take somewhat longer. Indeed it will be a task that occupies most of the 2030s. By this time there will be considerable international cooperation involving the European Space Agency and others. This will be advantageous to the US as they will still lead the project and will be able to rely on partners to develop costly aspects of it. It will also internationalise the project by treaty and make it harder for US administrations to cut it. But the experience and information gained from working on the Moon will not just concern the Moon, as the first voyages to Mars will be taking place during that decade. From the mid-2020s as the Mars voyages are planned, what we do on the Moon and what we plan for Mars will be intertwined, for as we shall see there is ice on Mars and travellers there will be in far greater need of it than those going to the Moon, with their relatively swift escape routes back to Earth.

There is an obvious brute force method of extraction. Use a tractor to collect the surface material and deliver it to a smelting unit that will heat it, allowing metals to segregate and the ice to sublimate (turn from ice directly into water vapour). The vapour can then be condensed on a cold surface to produce water, which is duly collected. Of course, building, operating and maintaining such equipment in such an extreme environment would be difficult.

The nature of the regolith is a severe problem. It's composed of shards of very strong material and is very abrasive. It also gets everywhere – just look at pictures of the Apollo moonwalkers back in the lunar module after their excursions, when they were covered in dust. Additionally, essential parts of robotic vehicles, such as joints, bearings and motors, may not work well in the extreme cold of the Moon's dark craters. Corrosion and lubrication will need to be investigated. Metal tyres will wear out quickly, as has been seen on the Curiosity rover on Mars.

But can the water be extracted from the regolith without heavy lifting equipment? One method that could be tried on early Artemis missions is using the smaller craters on the rim of Shackleton. They could be covered with a small flexible tent or a transparent gel-like substance. This would cause the regolith to heat, allowing the capture of the water vapour.

The first forays into Shackleton crater itself will take a great deal of planning. Perhaps a smaller crater than Shackleton will be used by astronauts to rappel into, to dig around for the ice themselves. It might be in the form of a dirty snowbank, or ice mixed with sulphur oxides and ammonia, and all combined with shards of regolith and glass.

By Artemis mission 5 it will be time to take a look inside Shackleton and the first step will be sending something over the lip of the crater to the upper reaches of its slopes to measure the strength of the surface. I imagine an uncrewed rover moving slowly to the edge. It would deploy a smaller rover configured like an instrumented axle with wheels on either end that would be lowered down the side. Lights and cameras will show its progress, either by microwave link or a power and data umbilical. It could be hauled up and lowered from a different position at a later date. Samples would be taken, and their water content analysed. By this time scientists will have an accurate three-dimensional understanding of the crater's side and floor thanks to infra-red lidar technology. They will also have a temperature map of the depths. The temperature in the bottom is the reason why there is ice and it will present great challenges. It's much colder in these craters than it is on the surface of Mars. It might be worth developing an elevated track system for pallets to move around and into and out of the crater and perhaps later on a sintered highway.

By the mid-2030s we may reach the crater floor after using try-outs of many different designs of equipment and having placed some processing facility there – not a full-scale one but enough for a proof of concept and to work out teething problems. Mirrors and then heliostats could be placed on the sunlit rim to illuminate the crater and a so-called power tower of solar panels erected to gather solar power. Some of the light from the heliostats would be reflected into the processing plant and used to sublimate the ice. To split water into hydrogen and oxygen requires a lot of energy and estimates for the total energy for this and transport and storage come in at about 3 megawatts. This can come from solar panels or a nuclear reactor. There are advantages and disadvantages with either method. Once liquid hydrogen and oxygen have been made there is the non-trivial question of transport and storage; one can imagine a tanker moving between production site and storage tanks. Even in the cold of the crater, hydrogen, with its 20.28 K boiling point, will be hard to keep from boiling.

The more we experiment on lunar regolith the more options we will have. A group of European scientists has recently found a way of obtaining oxygen by electrolysing a simulated lunar regolith. A prototype of a device that can do this has worked well. It's made by the British company Metalysis who have reported their findings in the journal *Planetary and Space Science*. There are plans to produce a version of it that could be taken to the Moon for testing. When the oxygen has been extracted from the regolith, what's left – a mixture of iron and aluminium and other materials such as titanium, magnesium, silicon and calcium – could prove useful, for example, for making bricks or as material to be fed into a 3D printer.

When it comes to lunar resources other than water, many are excited by the possibility of extracting an isotope of helium – helium-3 – from the lunar regolith. It's present in tiny amounts, but if it's possible to gather it in significant quantities it may prove useful in the future by providing energy back on Earth in a fusion reactor.

Energy from nuclear fusion – the fusing of atoms – rather than reactors in use today that use nuclear fission – splitting atoms – has been a dream for the whole of my life and we are still some distance away from a working fusion reactor, let alone a commercial one. But if one could be made it could change our energy outlook significantly. At present, the world's fusion programs are investigating the easiest fusion reaction, which combines deuterium (hydrogen with a nucleus consisting of one proton and one neutron) and tritium (hydrogen with a nucleus consisting of one proton and two neutrons). Deuterium is not radioactive and occurs naturally on Earth in low quantities. But if you can gather enough to 'burn' it in a fusion reactor it could release a vast amount of energy. The other component to be used, tritium, is mildly radioactive and decays, so it is made in another design of fusion reactor.

Helium-3 is in moon dust because the stream of particles given off by the Sun contains small quantities of it and over geological time it has been implanted in the regolith at about four parts per billion. That's very little. To get a useful quantity – a few kilograms – you would need to process over a million tonnes of regolith. A kilogram of helium-3 if burned in a fusion reactor could possibly produce 100 million kilowatt-hours of electricity. Unfortunately, fusion reactors that burn helium-3 require significantly higher temperatures to work than the fusion reactors we are currently contemplating, and they would be extremely difficult to make.

The desire for helium-3 fusion energy is a noble one, but the figures let such dreamers down. You would have to excavate a volume of regolith kilometres long on each side and ten centimetres deep. All this material would have to be heated to about 700°C, causing the helium and all other volatiles to be released. If you could process a tonne every minute, it would take at least 150 days. Then you would have to separate the helium-3 from the far more abundant helium-4. (You would get 1 kilogram of helium-3 for every 2,500 kilograms of helium-4.) This could be achieved by cooling it to very low temperatures, where it will fractionate because the two isotopes have different boiling points. In the process nitrogen, hydrogen, and carbon would be released, which would be very useful for a lunar colony. Given all this, exploiting helium-3 is a long way off and certainly not something that will happen within the next 50 years. But if we do ultimately achieve it, lunar helium-3 could power human civilisation at its current level of energy consumption for about a thousand years.

So for the near future, water, with its potential to be turned into liquid oxygen and liquid hydrogen, remains the key lunar resource for powering spacecraft. Travelling through the solar system is all about the economics of energy. By far the most energy used on any space journey is in the initial step from the Earth's surface to low Earth orbit – just

300 kilometres or so. Getting out of the Earth's gravitational well is hard. Someone once said that just getting into low Earth orbit gets you halfway to the Moon. Actually, it gets you halfway to anywhere in the solar system. But a closer look at the details of using a rocket fuel produced on the Moon to fuel a spacecraft to go to Mars shows that in the majority of cases there might not be any benefits if the Mars spacecraft were launched from Earth as the energy needed to go from low Earth orbit to Mars is less than that needed to get rocket fuel from the Moon's surface to low Earth orbit.

The first missions to Shackleton will use an unpressurised lunar rover for short trips but for longer trips of about 20 kilometres or so a pressurised cabin will be required. Toyota, along with Japan's national space agency, is coming up with a design. In artist's impressions it looks a bit like a big off-world SUV. 'Manned, pressurised rovers will be an important element supporting human lunar exploration, which we envision will take place in the 2030s,' Koichi Wakata, vice-president of the Japan Aerospace Exploration Agency (JAXA), has said. JAXA aims to launch the new rover in 2029. It will be about 6 metres long and 5 metres wide and will feature a 4-cubic-metre cabin capable of accommodating two passengers – four in an emergency. It will be powered by fuel cell technology similar to what's used in some of Toyota's Earthbound vehicles. Fuel cells have been used in space since the 1960s and run on oxygen and hydrogen and emit only water. According to Toyota the rover will have a range of more than 10,000 kilometres, a vast improvement over the lunar buggies used by the Apollo 15, 16 and 17 missions.

Apollo 17 astronauts Gene Cernan and Jack Schmitt set the distance record, driving their rover a total of 36 kilometres on three separate outings in December 1972.

The first missions for the pressurised rover will probably be driving around the rim of Shackleton, some 62 kilometres, and then along the ridges leading to the rims of nearby craters. In the 2040s the long-range exploration of the Moon will begin. The most exciting of these early trips will be to Mount Malapert and back, a round trip of about 350 kilometres. A longer trip that could be considered is from Shackleton to Schrödinger crater and back lasting about 90 days and covering about 1,100 kilometres. It is thought that Schrödinger might have mantle rocks excavated from a depth of 200 kilometres during the formation of the south polar Aitken basin by a giant impact billions of years ago. The baseline for this mission is four astronauts working in teams of two. They would take two rovers and a logistics vehicle. In many scenarios the rovers need to be recharged for 24 hours after three days of use. They will have a top speed of about 5 kilometres an hour and will be driven for eight hours a day. Earlier expeditions could place what are called Portable Utility Palates (PUPs) along the route. The PUPs would carry batteries that could be used to replace depleted ones in an emergency. Science packages could be left in Schrödinger basin. Other science activities include panoramic visual surveys, laying geophones for active seismic sensing, collecting interesting rocks, and acquiring regolith samples.

With such expeditions, the Moon's trackless deserts will be opened up as we will see sights we have only imagined.

We could work our way up the ramparts of Mount Huygens, which at about half the height of our Mount Everest is the Moon's tallest mountain; drive along the 166 kilometres of the strange so-called Alpine Valley that radiates from the north-east section of Mare Imbrium and follow the rille that runs along its middle; climb the shallow slopes of Mount Pico, an isolated mountain peak in Mare Imbrium that casts long shadows over the plains that surround it, perhaps thinking about the space battle Arthur C. Clarke set there in his novel *Earthlight.*

We have only begun to appreciate the sights the Moon has to offer. There are other aspects of the Moon we have known only too well for decades.

THE BALANCE OF
POWER ON EARTH

———•———

In May 1961, just days after President Kennedy told the American people of his goal of landing on the Moon before the end of the decade, the US Air Force finished a secret report on military plans to colonise the Moon.

They had in mind a sort of Fort Luna which would consist of pressurised empty fuel tanks 3 metres in diameter by 4 metres long buried into the lunar regolith. They would be used for barracks, airlocks, a hospital, a laboratory, a command post and a mess room. On the perimeter of the base would be the arsenal housing explosives and munitions. Four small nuclear reactors would provide electricity. This, and the idea of exploding a nuclear bomb on the Moon to demonstrate US power, was among many military proposals for its occupation. The USAF believed that the Moon was the ultimate high ground in the battle for strategic supremacy, the 'fortress of the next conqueror of the Earth', as one report put it. *Collier's*, the magazine that Wernher von Braun had used to set out his grand vision of lunar exploration, once ran an article, 'Rocket Blitz from the Moon', showing V-2-like missiles emerging from lunar silos. For some in the US military the Apollo Moon landings were only part of a much bigger picture, a life-or-death one.

One of those who passionately believed this was the USAF's General Homer A. Boushey, the director of the USAF Office of Advanced Technology. In a lecture to Washington's Aero Club, he outlined two military uses of the Moon, as a missile base and as the place from which to spy on the Russians. He said that missiles fired from the Moon could be guided from start to impact and that a Moon base would be almost invulnerable. Any attack on the United States would be easily seen from the Moon, and 'sure and massive retaliation' would follow. If the Soviets were to attempt to destroy the Moon base, they would have to fire missiles towards the Moon two-and-a-half days before they attacked the United States. Boushey said, 'We cannot afford to come out second in a territorial race of this magnitude. This outpost, under our control, would be the best possible guarantee that all of space will indeed be preserved for the peaceful purposes of man.' He added, 'It has been said that "he who controls the Moon controls the Earth". Our planners must carefully evaluate this statement for, if true – and I, for one, think it is – then the U.S. must control the Moon.' Remember those words.

General Dwight Black, the USAF director of guided missiles and special weapons, told Congress, 'I would hate to think that the Russians got to the Moon first. The first nation that does will probably have a tremendous military advantage over any potential enemy.' The United States Army also had its eye on the Moon: they listed their requirements as a manned lunar outpost and a lunar tank. Another study was carried out for the ballistic missile division. It said, 'the lunar base possesses strategic value by providing a site where future military deterrent forces could be located', and that 'a military lunar system has potential to increase our deterrent capability by insuring positive retaliation'. It added that such a base could become operational in June 1969.

The Army Ordnance Missile Command called its study 'Project Horizon'. It was produced by a group headed by none other than Wernher von Braun and it was characteristically large in scale. It said a lunar outpost was 'required to develop and protect potential United States interests on the Moon, to develop Moon-based surveillance of the Earth'. It added that an outpost was of such importance that it 'should be a special project having an authority and priority similar to the Manhattan Project'. It warned that the Soviet Union had openly announced that some of its citizens would celebrate the 50th anniversary of the October 1917 revolution on the Moon. If the Soviets were to be the first to establish a Moon base, it 'would be disastrous to our nation's prestige and in turn to our democratic philosophy'. It concluded chillingly: 'Throughout the recorded history of human endeavour, the military outpost has been the hub around which evolved the social, economic, and political structure of civilisation. During early U.S. history, the establishment and maintenance of routes of communication to the far west was made possible by Army outposts.' The wild west had come to the Moon.

But President Eisenhower, with his growing fear of the military-industrial complex, as he described it, was having none of this. He was right in dismissing the military significance of Sputnik, even if few of the American people could see it. Few of the apocalyptic descriptions of the military significance of space have come true. As far as lobbing nuclear bombs onto the Soviet Union was concerned, a satellite was not the best way to do it. A satellite was vulnerable and not all that accurate upon re-entry into the Earth's atmosphere. Intercontinental ballistic missiles were far more convenient and reliable. Nor was there any real military advantage in having a base on the Moon. It was too visible and too far away. Missiles from it would take over two days to reach Earth, during which time they could be intercepted. It was, according to one report, a 'clumsy and ineffective way of doing a job'. To his credit, Eisenhower set the tone for future manned space endeavours when he awarded the responsibility for manned, Earth-orbiting space flight to the newly created civilian NASA. Since then, military space professionals have historically only considered the space from geostationary orbit down to Earth.

But that is changing. Today the language is returning to that of Homer Boushey and Dwight Black.

Today there is increasing military interest in cislunar space – the region extending beyond Earth to the Moon – but this time it's not the Russians that worry America, it's the Chinese. Both countries realise that cislunar space is strategically vital because the future exploitation of space resources has the potential to alter the balance of power on Earth. It

will be an area of competition, tension and, inevitably sometime in the coming decades, conflict. Cislunar trade routes
and lines of communication will need protecting, especially
as many see it as the 'high ground' – a position of advantage
or superiority. Today the US military sees virtually everything China does in space as a threat and this is leading
towards the weaponisation of space. Simply put, cislunar
space is a new flank.

Peter Garretson leads the USAF's Space Horizons
Research team; he points out that China's Chang'e 4 far side
mission and its Magpie Bridge relay satellite at the Earth–
Moon L2 gravitational balance point are part of a long-term
plan. 'They have already put in place the first node in a
broader communications architecture, and perhaps a cislunar space domain awareness system as well,' Garretson said.
'Next comes sample return, polar landings and 3D printing
of a "Lunar Palace" with an industrial mission to make economic use of lunar resources.' Where China takes an interest
the US defence community will follow.

For some the threat cislunar space poses was brought
home quite a while ago, in 1998. On Christmas Day 1997,
Asiasat-3 was launched from Russia by a Proton rocket.
The rocket's fourth stage was to give it a final kick into the
right orbit, but it burned for only two out of the planned
110 seconds, leaving the valuable communication satellite in
a useless orbit. For a while it seemed that it was a write-off
but then satellite operators came up with an audacious plan:
they could use the satellite's reserve fuel to swing it by the
Moon, using lunar gravity to alter its trajectory and bring the

satellite down to its originally intended orbit. In the initial plan this involved the satellite approaching the Earth from an unusual, usually ignored, direction where it was impossible to track. It had inadvertently become an invisible satellite in what was then described as a stealth orbit. It was an eye-opener for military space planners, showing to them how difficult it was to survey cislunar space, and China knows the implications of this very well. Their spacecraft Chang'e 2 was in lunar orbit for a year before it was moved into a large so-called halo orbit around the Moon where it stayed at a gravitational balance point for eight months before being dispatched for a rendezvous with the asteroid Toutatis. It was an impressive feat, showing they can place satellites anywhere in cislunar space.

Cislunar space offers a vast area in which to perform manoeuvres: one that is difficult to survey and from which surprises can then emerge. It's analogous to deep-sea submarine warfare. It's clear that China's military-run space program is positioning itself in cislunar space. 'We are behind, and we must catch up,' Garretson said. 'What is driving the US military to look at cislunar is not some present tactical advantage,' he added, 'It is fear that China's moves to cislunar space will provide it with a positional and logistic advantage from which it could occupy, constrict, threaten or coerce US interests.'

A recent information note issued by the United States Air Force (before the Space Command took over) summarises their thinking. 'As the space beyond geosynchronous orbit becomes more crowded and competitive, it is important for the Air Force to extend its space domain awareness responsibilities to include this new regime,' says the pre-solicitation. 'The Air Force is seeking commercial innovation in support of space domain awareness for future cislunar operations.' Some believe that without a plan to industrialise the Moon, the USA will find themselves confronting a juggernaut with an industrial, logistical and manoeuvring advantage.

Both sides know it. Recently a senior Chinese general had a warning for the US regarding a future arms race in space: be prepared to lose. Qiao Liang, a major general in the Chinese air force who co-wrote a book called *Unrestricted Warfare: China's Master Plan to Destroy America*, said in an interview, 'If the United States thinks it can also drag China into an arms race and take down China as it did with the Soviets … in the end, probably it would not be China who is down on the ground.' He added that outspending a rival power into economic exhaustion might have helped the US win the Cold War, 'but it won't work against a wealthy manufacturing powerhouse like China, China is not the Soviet Union.'

Many hope that today's attitudes will not set a precedent for decades to come but at present neither country seems interested in a diplomatic route to lower the tensions; there is little dialogue between the US and China. The Wolf Amendment, first passed in 2011, forbids the US government from working with China and prohibits any cooperation between the China National Space Administration and NASA. China is barred from participation with the space station.

The threat is that we will over the next few decades in space build up the technology and tension that could put the

US and China in a position for space conflict. Qiao Liang said China is not seeking a space war. But he said it is preparing to counter any nation, including the US, that seeks to pose a threat to China's national security. 'When the United States and the Soviet Union engaged in the Cold War and the arms race, the United States was the largest manufacturing country, and the Soviet Union was not even the second,' he said. 'But today it is China who is the world's top manufacturer.'

China's efforts are advancing quickly and there is some evidence that their weapons can already attack targets much farther from the Earth than can those of the United States. China is believed to have the capacity to use missiles to attack satellites in geosynchronous orbit, or 36,000 kilometres above Earth. Its reliance on space assets is also increasing: it has more than 120 intelligence, surveillance and reconnaissance satellites – second only to the United States. It is also investing in the development of several anti-satellite missile systems as well as developing satellites that could be used for repairs in orbit, or to disable a satellite. A Pentagon China Military Power report found it is pursuing jamming and 'directed energy' weapons that can disable and destroy satellites to 'blind and deafen the enemy'.

Some believe China will soon have a ground-based laser that can blind optical sensors on satellites in low-Earth orbit. This is a gamechanger because a strike in geosynchronous orbit would produce a debris field that would render this critical region unusable, disabling its use for missile warnings, weather observations, TV and communications. According to Kaitlyn Johnson, an associate fellow who specialises in space security at the Center for Strategic and International Studies, the USA has 'much more to lose in GEO [Geostationary Earth Orbit] than any other country'.

The Chinese government insists that it is merely responding to aggressive US moves to dominate space militarily. Qiao Liang called it 'bullying and hegemonic' for the United States to insist that other countries can't follow suit: 'China's purpose to develop space capabilities, firstly, is we do not want to be blackmailed by others ... we hope to use space peacefully. But if others want to oppress us by occupying the heights of space and opening up a "fourth battlefield" China will certainly not accept it ... China is developing many of the same space capabilities the U.S. did decades ago, while the U.S. is focused on sustaining its capabilities and making them more resilient,' he said. 'On the whole, the U.S. is still far more capable than China is, but the relative advantage is narrowing.'

The US military are responding by accelerating the development of advanced technologies that can be used in cislunar space, such as robotic servicing satellites, and they want a new nuclear-fuelled rocket for operations between Earth and the Moon – a region which they say is 'in danger of being defined by the adversary'. They are also planning what they call 'ubiquitous satellite command, control and operations'. They say, 'We are seeking technologies that enable rapid, ubiquitous command and control of satellites.'

By the early 2030s the Moon base will be in its infancy, taking tentative steps towards the production of water. For some that will be enough but colonisation will not satisfy some when there are new conquests to be had. As the footprints spread on the Moon, the attention will increasingly turn to Mars. After the 2030s the Moon will still have a rich and surprising future, but Mars will be the frontier.

Mars is not like the Earth or the Moon, especially not like the Moon. It has craters and canyons, mountain ranges and deserts with shifting sand dunes, but they are more than analogues of similar features on Earth. Mars has made its own world using things it has in common with the Earth. You could look down on the magnificent Valley of the Mariners and watch the morning mist clear as the Sun rises. There are ancient shorelines and dried-up river beds, extinct volcanoes and possibly some that are just dormant. A peach-coloured sky with clouds, and possibly running water; and ice, loads of ice. The Moon is an outrigger world, Mars is the first true world of the Cosmos.

From now on the story of the Moon and Mars must be told together. But first, a warning from Apollo.

The Apollo astronauts did not stay in space for very long, a few days above the protective magnetosphere at most, but it had its effects. One study shows that Apollo astronauts had high mortality rates due to heart disease. The study compared the mortality rates of Apollo astronauts to those of astronauts that lived on the International Space Station. They found the occurrence of cardiovascular disease-related deaths among those who travelled to the Moon was higher. The rate among astronauts who never flew is 9 per cent. Among low-Earth orbiting astronauts, it's 11 per cent, but for those who went to the Moon it is 43 per cent, four to five times higher. This is despite astronauts having a higher quality of life than most, their incomes are relatively high, they are physically fit, and they have access to good medical care. That should confer on them a lower incidence of heart problems, but it didn't. Researchers also exposed mice to simulated space radiation and after six months they had damaged arteries.

This new research, and other studies of a similar nature, casts a long shadow over missions to Mars. You cannot regard going to Mars like going to the Moon and continuing a bit further. It is more fundamental than that.

PART 2

TO MARS

I love all waste
And solitary places; where we taste
The pleasures of believing what we see
Is boundless, as we wish our souls to be
[…]
Meanwhile the sun paused ere it should alight,
Over the horizon of the mountains; Oh,
How beautiful is sunset, when the glow
Of Heaven descends upon a land like thee,
Thou Paradise of exiles …

—PERCY BYSSHE SHELLEY,
'JULIAN AND MADDALO'

WHEELS AND SHADOWS

———•———

Valles Marineris.

Mars.

2069.

By 2069 humanity will have survived on Mars for 6,750 Martian days, or sols, continuously. For everyone who has been a part of it Mars has either become more beautiful or more of a struggle with each passing day. There are those back on Earth who lived for many years on Mars and in between the planets. Having spent so long in trackless deserts and empty plains of Mars, the Earth is overwhelming. Each breath a reminder of just how alien is the red planet, and how it never leaves your soul.

For many new Martians, the Sun is an unnerving sight, strangely shrunken, with no warmth as far as it's possible to tell from inside a spacesuit or habitat. There are two hours of Martian daylight remaining. The shadows are lengthening as night-time creeps over Xanthe Terra – the Golden-Yellow

Land – and the great outwash plains that merge into Chryse Planitia – the Plain of Gold – and approaches the eastern canyons of the Valley of the Mariners, overwhelming the depths of Capri and Eos Chasma. The vehicle's shadow is spreading over a terrain similar to Earth's Gobi Desert: dust scattered over everything, brown and pink stains against ochre, rocky outcrops, blocky seams, rippled terrain, gravel fields, a hint of stone ramparts on the horizon. The region is termed geologically complex, which in this case means that it has seen fire, ice and water, though that was billions of years ago. Signs of ancient water are everywhere, but the landscape is dead: only wheels and shadows move.

It's a twenty-minute drive back to the base and excursion rules say it's time to return. Although the base is within sight, there are illuminated waypoint markers and the rover with its human cargo makes its way automatically. They leave behind the ice drill they were using to determine ice depth in the region to be explored for resources. The drill was being used to confirm the data from the ground-penetrating radar that indicated the ice was 20 metres down. It was essential that the colony gain access to it, but just you try drilling on Mars.

To the rover's left, several hundred metres away, is a region cordoned off with blue markers. That was off-limits to almost everyone, a region that had been identified years earlier from orbit as an area where life could possibly have hung on for millions if not billions of years. A special protocol was required when approaching that area.

As the rover reached the habitat the last rays of sun were glinting on the solar panel fields and the vegetation modules.

The temperature was declining rapidly. There was to be a celebration tonight.

In human experience there will never be an isolation as total as that of Mars. If its visitors stepped outside the habitat modules they would find nothing in the way of comfort. No movement except for the little caused by the wind and the rare landslide. No sound save for their own breathing in their pressure suit. If any of them were to climb the nearest hill, all they would see would be more hills and desert plains, an endless vista for the next 21,344 kilometres until they arrived back at their base. The circumference of the Earth is almost double that of Mars but because the Earth is mostly oceanic, one could travel further in a straight line on Mars than on Earth, and find no one. Travel around the red planet and you may find the wreckage of a spacecraft or a worn-out rover, but for the most part there is nothing human there.

Just like to the explorers of the heroic age of polar exploration 150 years ago, celebrations were important for morale. They were vital for crew overwintering in Antarctica. At midwinter in the extreme dark and cold they would open their bottles of claret, tinned anchovies and corned beef as they sat around their stoves. They thought little of the fire that kept them warm and nothing of the air they were breathing. That was the constant backdrop of their environment. On Mars there is no such backdrop. Every breath has to be measured, every calorie counted, every scrap of energy generated and accounted for, every crew member and relationship tolerated, every second of life planned for and appreciated. There is no claret on Mars.

As the Martians go about their allotted tasks, in the back of everyone's mind is the unacknowledged anxiety surrounding the fragility of living on Mars. The experienced Apollo 12 moonwalker Alan Bean admitted to this feeling after his Apollo and Skylab missions: 'I felt real fear out there, particularly when thinking of dangers such as window failure and depressurisation' – and he was only at most three days from Earth, not 300. Martians like to say that no one can truly appreciate how far it is from Earth, in distance and emotionally, unless one has been there.

Despite what had been optimistically predicted 50 years previously, there is no extended infrastructure on the red planet. No spread of outposts in interesting places, no Martian economy. The assumption that sending humans to Mars would proceed systematically and swiftly once we had returned to the Moon was too optimistic. Of course it was going to take longer than the plans suggested. Going to Mars was nothing like going to the Moon. Mars is not the Moon. Whoever said it was like the Moon? Human Mars missions were done more slowly. They were more costly than those who saw them as an extension of going to the Moon imagined, and they tested humanity far more severely. Life on Mars depended on the supply route from Earth, for the colony was not self-sufficient. Plants had been coaxed from the Martian regolith, but it was as yet just a supplement. People could not live on Mars without the Earth.

The celebrations of 2069 are a time for the colony to look back on how they got there, assess where it is going and ask if mere survival is enough, or enough for now? This was no easy migration, nothing like the science fictions of the 2020s when billionaires talked of tens of thousands embarking for Mars within years and of a million people living there by 2050. The absurd dreams of a vast migration to start again on a new world collided with the harsh reality of off-Earth living. Mars was hard won, and it is too soon to tell if any victory there has been. Some of the colonists of 2069 cannot return home. There are many graves on the red planet and in the space along the colonists' route. Yes, the red sand had footprints, but because of the nature of Mars they will not last as long as those on the Moon, for any number of reasons.

When will we go to Mars? When will we land on its red, magnetic dust? These are questions we cannot answer, but it seems that we will not do so in the next decade, or even, perhaps, in the decade after that, despite what some billionaires will tell you. It is already too late to launch a human mission before the late 2030s, despite the acceleration of our return to the Moon. A recent report prepared by the Science and Technology Policy Institute in the United States and NASA noted that in 2017 the US House of Congress put a line in NASA's budget for a 'Mars human space flight mission to be launched in 2033'. The report considered this and concluded, 'We find that even without budget constraints, a Mars 2033 orbital mission cannot be realistically scheduled under NASA's current and notional plans.' It continues, 'Our analysis suggests that a Mars orbital mission could be carried out no earlier than the 2037 orbital window without accepting large technology development, schedule delay, cost overrun, and budget shortfall risks.' The next launch window, in 2035, was also deemed infeasible, pushing the earliest possible date for flying the mission to the following launch window in 2037. As I said, going to Mars is not like going to the Moon, and then carrying on a bit further.

Some have argued that the Trump initiative to put humans on the Moon by 2024 could bring a human Mars mission closer. The head of NASA Jim Bridenstine has argued that going to the Moon in 2024 would allow an earlier human mission to Mars, adding, 'People say, "Why are you accelerating a mission to the Moon?" Well, because it accelerates a mission to Mars.'

One person who is sure we will get to Mars a lot earlier, in fact within years, is Elon Musk, whose Moon plans we have discussed. As well as placing cargo on the Moon, even before the next human landing on the Moon he wants to 'send our first cargo mission to Mars in 2022. The objectives for the first mission will be to confirm water resources, identify hazards, and put in place initial power, mining, and life support infrastructure.' As if that wasn't ambitious enough, he has in mind a second mission, with both cargo and crew, targeted for 2024, which will construct a propellant depot and prepare for future human missions. The ships from these missions will serve as the beginnings of a Mars base, then will come a thriving city and eventually a self-sustaining civilisation on Mars. In a speech to the US Air Force, he said its stated launch cost of about $2 million per mission should allow a 'self-sustaining city on Mars'. To do this SpaceX will need to build and fly around 1,000 Starships that will transport cargo, infrastructure and crew to Mars over the course of around 20 years. In addition to its Mars objectives, Musk believes that with the Starship SpaceX will be able to launch upwards of 10 million tons to orbit per year.

This is not realistic. SpaceX have got carried away with a dream and will not be able to turn imagination into reality. I've seen such overly wishful thinking before. It's easy to get carried away. When I was a young space scientist I wrote for *The Times* newspaper and I thought they were generally immune when it came to the outrageous space stories that the tabloid newspapers frequently ran. You know the type of thing, the Martian mountain that looks like a face (under certain resolutions and illuminations) being evidence of an ancient civilisation, alien spaceships stored in Area 51 of the Nevada US military test and training range. But in 1987 they fell for one of the most outrageously hyped space stories I have ever come across.

Russia, it seemed, had a masterplan for the industrial domination of space. They were planning kilometre-wide mirrors in space to light cities and boost crops in the 1990s. A later phase, planned for 2002 would see giant solar-cell power stations, sending hundreds of megawatts of power down infrared beams to receiving stations on Earth. One expert said, 'I am convinced from studying Russian space research published by many of their leading experts that the Russians are now years ahead along the path to space industrialisation and poised to gain benefits which would give them economic leadership of the world.' But it was unrealistic. To achieve their aims they would have had to launch their super rocket, Energia, capable of lofting 270 tonnes, every eight hours, indefinitely. That means they would launch 295,000 tonnes into space every year. One of the major problems was that even in 1987 it was obvious that Energia was not going to be an operational rocket; it only flew twice. How does 295,000 tonnes a year compare to Elon Musk's 10 million tons a year? My point is that there are dreams and there is reality. The purported Russian plans for the industrial domination of space in the 1980s obviously went against what was feasible, and so it is the case with sending thousands of Starships to Mars in the next few years, or even the next 50.

In a sense, Musk's space rival is the richest man alive, Jeff Bezos, who founded the aerospace company Blue Origin in 2000. A test flight of the rocket Blue Origin is developing reached space in 2015 and he plans, like Virgin Galactic, to begin commercial suborbital spaceflight. Blue Origin has also designed a lunar lander that NASA has chosen to land on the Moon. Bezos doesn't think humans will ever want to live on Mars. 'My friends who want to move to Mars? I say, "Do me a favour, go live on the top of Mount Everest for a year first, and see if you like it – because it's a garden paradise compared to Mars."' He imagines humans living in self-sufficient space structures, like the huge cylinders designed by Princeton physics professor Gerard O'Neill forty years ago. Jeff Bezos once said that space is really easy to overhype. He's right.

Mars presents unique challenges. We have to learn about the planet, and we have to learn about ourselves.

As I envisage it, the first people to reach Mars will not land there. After a 201-day voyage, the first ship – I will call it the James Caird II – enters the orbit of Mars in 2039. This is several years later than had once been proposed and not landing is controversial. Why go all the way to Mars and face all the risks and not land on its surface, say some. But there are good reasons.

The voyage of the original James Caird was one of the most dangerous and dramatic of any voyage undertaken by humans. Thirteen hundred kilometres through the roughest seas on Earth, between Elephant Island in the South Shetland Islands to South Georgia. An exhausted crew of three, including Ernest Shackleton, battled through the southern seas and the 'furious fifties' with their hurricane-force winds and giant waves in a small boat. That they reached South Georgia was a miracle and after landing on the island's west side they had to cross the mountainous interior to reach the whaling station from where, eventually, they were able to rescue the crew they had left behind on Elephant Island. Of the voyage, navigator Frank Worsley said, 'We knew it would be the hardest thing we had ever undertaken, for the Antarctic winter had set in, and we were about to cross one of the worst seas in the world.' The worst sea in the world is a pleasant cruise to what a voyage to Mars can do to you.

It's early spring on Mars and the crew are waiting for the global dust storm to abate, just as scientists several generations earlier had to wait when another spacecraft arrived during a dust storm. It will keep them waiting in orbit for longer than intended. If they are to make the trip back home without having to stay on Mars for another 200 days, they will have to leave in 32 days. Provisions have been made for such a contingency, but it would be a desperate voyage.

As the dust storm clears, the crew of the James Caird II become the first humans to look directly on the magnificent sights of the red planet. In a way the first to experience the emotional impact of Mars. As I have said, thinking of Mars as a different type of Earth, possibly an Earth in waiting, is wrong, and likely to get one killed. Mars is Mars, its features are Martian, forever its own.

No planet has embedded itself so deep into our culture as Mars. It must have started on some far-distant unmarked African night when our ancestors, not yet human but with a spark of comprehension, pondered on its strange yet familiar colour. It became the fire star to some, Horus the red to the ancient Egyptians, the blushing star to others. But to many it was a like a drop of blood in the sky. Mars was war and ruin, pain and death.

The crew of the James Caird II are of course far from the first to look up and wonder about Mars. Bernard Le Bovier de Fontenelle (1657–1757) was a great French writer in the Age of Enlightenment and for his work he has a crater on the Moon named after him. But he was unimpressed by Mars: 'Mars hath nothing curious that I know of, his Day is not quite an hour longer than ours, but his Year is twice as much as our Year; he is a little less than the Earth; and the Sun seems not altogether so large and so bright to him, as it appears to us; but let us leave Mars, he is not worth our stay.' As telescopes became better, Mars became more interesting and more than a symbol or the fourth planet from the Sun – a small yet curious red disc.

Giovanni Virginio Schiaparelli's father was a furnace maker who in 1839 took his four-year-old son outside one night to look at the stars through the clear skies of the foot-hills of the Alps. Giovanni went on to become the greatest observer of Mars in the 19th century, and he was very excited by the favourable position of Mars for viewing in 1877. Schiaparelli prepared for it obsessively, avoiding 'everything which could affect the nervous system, from narcotics to alcohol, and especially ... coffee, which I found to be exceedingly prejudicial to the accuracy of observation'.

Aged 42, he observed Mars during what was called the great opposition of 1877, when in September Mars was just 56 million kilometres distant. Opposition occurs every two years and is the time when the Earth and Mars come closest. Some oppositions are better than others: 1877 was a close one. Schiaparelli produced a map which he improved on during the subsequent opposition in 1879. But to other observers his map was baffling. They had seen hazy shades and little detail on the disc – but not so Schiaparelli. He saw streaks which he called *canali*. The Italian word means 'channels', and does not imply an artificial origin. However, others swiftly translated the term as 'canals' and soon speculation was rife about a dying planet whose inhabitants constructed giant canals to bring water from the polar caps to the equatorial deserts.

Also watching Mars that year was the American Percival Lowell, and he became obsessed by the canals and the possibility of life on Mars. Born into a wealthy Bostonian family, Lowell was an astronomer with a consuming passion for Mars. On a trip to Japan in 1892 he heard that Schiaparelli had been forced to discontinue observing Mars due to failing eyesight. Lowell decided to take up the work, but first he had to build an observatory. To Victorian astronomers it seemed obvious that it was another kind of Earth. It had an atmosphere and markings on its surface, including what seemed to be polar ice caps. Its day was only 41 minutes longer than ours. It was suggested we might try to signal to the Martians. One proposal was to plant trees in the desert in the shape of a right-angled triangle, proclaiming 'artificiality' over 56 million kilometres.

The 1892 opposition was the most favourable since that of 1877. Astronomer Andrew Douglass was in Peru and determined to see the canals. Through the clear Peruvian air, they appeared much as they did to Schiaparelli. But Douglass saw more: spots at the intersection of canals; and he saw lakes.

Back in America Douglass was trying to turn his observations into maps. Lowell asked him to find an ideal site for a Mars observatory. In a few months the observatory was built in Flagstaff, Arizona, standing on what was now christened Mars Hill. Between June and December 1894 Mars was observed almost every night by Lowell and by Douglass when he returned to Boston. In ten months, they made over 900 drawings, obtaining the crucial data for Lowell's theory of life on Mars. Lowell had abandoned the idea that the dark lines were canals per se, as to be visible they would have to have a width of 20–30 kilometres. He considered them to be wide paths of vegetation. Soon he concluded that they were part of a planet-wide irrigation network carrying water from the poles to the arid equatorial regions. The public were convinced that there was life on Mars thanks to Lowell's popular work *Mars* in 1895 and the 1897 publication of H.G. Wells's *The War of the Worlds*.

But in 1900 Douglass started conducting a series of experiments using discs or balls as 'artificial planets' which were observed through a telescope. At first Lowell encouraged him. Worryingly, lines occasionally appeared on a completely blank artificial planet, suggesting a peculiarity of human vision or telescopic imperfections. In 1901 Lowell observed such an artificial planet and drew a double canal where only light shading existed. Douglass was worried that the observatory's good work in many areas of astronomy were being put in jeopardy by the controversial Martian theories of Lowell. He wrote to Lowell's brother-in-law and business manager, W.L. Putnam, pointing out the low opinion held of Lowell's work by other astronomers. 'I fear it will not be possible to turn him into a scientific man,' he added. Putnam kept the letter from Lowell for four months. Suddenly, with no explanation, in July 1901, Douglass was dismissed. Douglass tried to be reinstated. 'I have at present all the assistants I want at Flagstaff,' came Lowell's reply. But the canals were in Lowell's imagination, not on Mars, and when he died, he was buried next to his observatory on Mars Hill.

There is one thing about Schiaparelli that Lowell and Douglass did not know and is not widely known even today. He had congenital eye defects, was colour blind, and had eyesight more sensitive to shades of grey and contrasts than normal. When he drew channels that were double – he called them *germinations* – few realised that they originated in his pathological eyes and not on Mars. In 1900, with fading eyesight, he said his farewell to observing. But he never gave up his belief in the channels. He knew it was an outlandish concept but on the top of a copy of one of his papers he sent to the famous French astronomer Camille Flammarion he wrote: 'Once a year it is permissible to act like a madman.' Shortly before his death he wrote to fellow Mars observer E.M. Antoniadi: 'The polygonations and germinations of which you show such horror ... are an established fact against which it is useless to protest.'

I wonder what Schiaparelli and Lowell would have thought had they been orbiting Mars in the James Caird II? Looking down with their maps in hand they would search for the grand Nilosyrtis canal that ought to flow northward from Syrtis Major, or the Heliconius twins traversing the edge of the north polar cap, and would look in vain for the giant centre of Trivium and the familiar equatorial Salaeus canal. But after their initial shock I like to think they would not be too disappointed at whey they saw.

From their orbit the crew of the James Caird II can see Aram Chaos, named after an ancient term for Syria, a heavily eroded crater on the eastern edge of the Valles Marineris and important because deposits of the mineral hematite have been found there. Dark terrain fills this ancient crater that was once a lake filled with sediments. Later the water froze and then melted, carving deep fractures in the sediments. Why Mars warmed after it had been frozen for over a billion years is a mystery. But water piled against the east wall of the crater and broke though, carving a gorge 80 kilometres long and 13 kilometres wide. The gorge is one of the wonders of

Mars, flowing directly into the Ares Valles outwash plain and onto where Pathfinder landed in 1997.

Olympus Mons was a variable bright patch to Schiaparelli and Lowell, but it is the largest volcano in the solar system. Schiaparelli called it Nix Olympia – the Snows of Olympus. Over 600 kilometres wide, it rises to 21 kilometres – three times the height of Mount Everest – with ramparts 1 kilometre high all around. Perhaps, it is not yet extinct but just lying dormant. To the east are three more volcanic peaks in a row: Arsia Mons, Pavonis Mons and Ascraeus Mons.

A short while later they will pass over the oldest region of Mars, the heavily cratered southern uplands that cover about a third of the planet. Schiaparelli chose the name Hellas, after the ancient name for Greece, for a giant impact basin. Its floor is 8.2 kilometres below the surrounding uplands and is riven by ancient drainage channels.

For curiosity's sake, one of the crew asks the computer to overlay on the surface the route of the most famous journey yet undertaken on Mars. In almost every class they have attended on Martian geology, footage of this journey has been used to show them what the red planet is like.

It was 3,200 kilometres long and took 57 days in a Mars rover with a top speed of 15 kilometres per hour. The journey was from Acidalia Planitia (named after the mythological fountain where the Four Graces were said to bathe) over a wide range of Martian terrain to the equatorial crater Schiaparelli.

If you were an explorer making this journey you would head south-east across Acidalia to Mawrth Vallis, one of the oldest valleys on Mars – a route between the southern uplands and the northern lowlands. Formed by catastrophic flooding, its light-coloured, clay-rich soils could be formed only in the presence of water. The region has always been high on lists of landing sites. The next section is a long climb into the outskirts of the densely cratered and heavily eroded ancient land of Arabia Terra, its numerous ancient canyons and valleys once carrying torrents to the lowlands. Passing mesas and pedestal craters – in which ejecta from craters protects the softer rock beneath it as the region around it is eroded away – you head into regions draped in layers formed by volcanic eruptions, reaching places formed by tectonic movements.

Next you have to traverse the 100-kilometre region between the craters Rutherford and Trouvelot and onto the crater Marth, one of the dustiest places on Mars. Thence turn south towards Meridiani Planum, a desolate region now, but 4 billion years ago a land of hot springs. Eventually you turn eastward and pass within 300 kilometres of the Opportunity rover, now long dead in the sands of Perseverance Valley next to Endeavour crater, travelling slightly south as you move across the equator to reach the north-west rim of Schiaparelli. Getting into the crater is one of the trickiest parts of the journey and has to be done by exploiting the dent in its walls made by Edom crater. A few more sols and you will have followed in Mark Watney's tyre tracks in the film *The Martian* as he reaches his Mars Ascent Vehicle.

The James Caird II continues its survey of Mars and leaves orbit after 29 days. Up to this point the mission has been a success.

TWENTY-TWO IMAGES

———•———

Just 22 images changed our romantic view of Mars for-ever. The fabled red world of canals, of lost civilisations, a fading, dying and drying world, a place of ancient cities crumbling to dust, dissolved into a handful of grainy images. The July 1965 Mariner 4 flyby showed us that there was no vegetation, no water, just dust and craters billions of years old.

They were murky, low-contrast pictures – some scientists suggested that the camera had a light leak – showing a sec-tion of the Amazonis desert near a dark patch called Trivium Charontis. The images cut a swath from Zephria towards Atlantis, Phaetonis and Memnonia before Mariner sped on, making measurements in interplanetary space before becom-ing another derelict probe orbiting the Sun. It was a difficult task getting the pictures back to Earth. They were taken by a TV camera and recorded on magnetic tape which could only be played back at a data rate of eight bits per second. It took weeks to send back data recorded in minutes. The images produced were just 200 pixels square.

Mariner also made observations of the Martian atmosphere. Two hours after it made its closest approach to the surface (9,850 kilometres), it passed behind the planet, at a point on the sunlit side between Electris and Mare Chronium. Its radio signal was distorted as it travelled through the thin atmosphere before being cut off. When it emerged at a point above Mare Acidalium the effect was seen again. From these observations the pressure of the atmosphere could be determined. It was low, very low: just 4 per cent of the pressure at sea level on Earth, and the atmosphere was 95 per cent carbon dioxide – unbreathable. It seemed likely that the polar caps were not made of water ice as had been supposed but of frozen carbon dioxide. Mariner 7 seemed to confirm this when it later passed over the south pole carrying instruments able to measure the surface temperature. It came out at $-123°C$, just what was expected for carbon dioxide ice.

The follow-on probes Mariners 6 and 7 sent back only 58 images between them, but they were of better quality than those from Mariner 4. Covering an equatorial region, and a part of the south polar cap, they increased the coverage to about 10 per cent of Mars. Mariner 6 had made some interesting observations of Hellas, showing it to be much smoother than the cratered terrain surrounding it. Jumbled ridges were also found, as well as terrain with no set pattern, described as chaotic.

The flyby results were fascinating but for many they were a disappointment. Mars was dull. The science that replaced the romantic Mars pointed to an inactive world of interest only to specialists. In many ways the flyby missions marked

President Lyndon B. Johnson and Vice-President Spiro Agnew
among the spectators at the launch of Apollo 11, July 1969.
NASA

The first man on the Moon:
Apollo 11 commander Neil Armstrong.
NASA

And the last to date: Gene Cernan after the final
moonwalk. Apollo 17, December 1972.

The Lunar Gateway will orbit the Moon, serving as a space
station and hub for lunar and deep space exploration.

A concept to explore dark lunar craters.
NASA

Exploring a lunar pit.
NASA

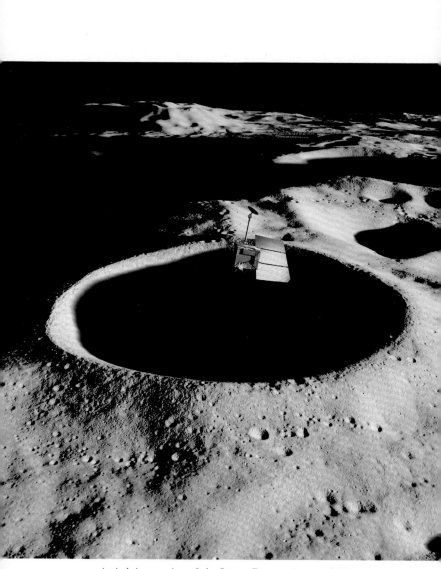

Artist's impression of the Lunar Reconnaissance Orbiter
over Shackleton crater.

NASA

A concept for constructing a lunar base using automated 3D printers, making use of lunar regolith as building material.

NASA

An artist's impression of an ice-processing facility with mirrors providing the energy.

James Vaughan

China's Chang'e 4 mission was the first to land on
the Moon's far side in January 2019.
CNSA/CLEP

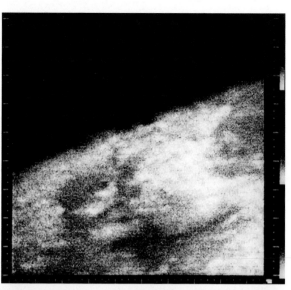

The first ever close-up picture of Mars, taken by Mariner 4 on 15 July 1965 and showing the boundary of Elysium Planitia. *NASA*

Mars, showing the Valles Marineris. *NASA*

Viking 2 landed in the Utopia Planitia region of Mars in September 1976.
NASA

Mars Pathfinder and its small rover Sojourner landed on Mars in July 1997.
It was the first successful lander since the Vikings in 1976.
NASA

The Curiosity rover on Mars. Composite image made of pictures taken by
the rover on 5 August 2015 after it had been on Mars for 1,065 Martian days.
NASA

Curiosity's tyre tracks looking back
as it drove over a sand dune in February 2014.
NASA

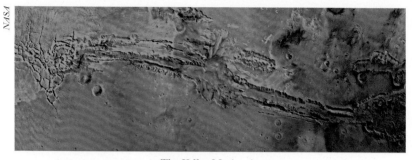

The Valles Marineris.

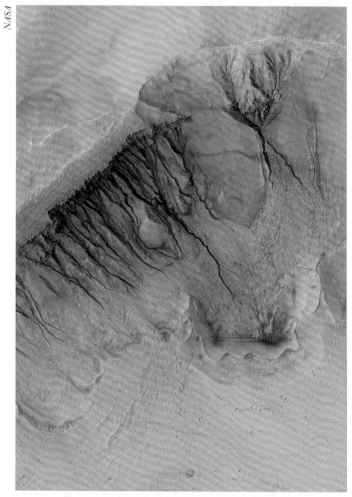

Recurrent slope lineae are thought to be caused by
melting sub-surface ice trickling downhill.

Artist's impression of a Mars transit spacecraft.

NASA

Speculative Mars habitats constructed by 3D printing methods.

Team SEArch+/Apis Cor

Scott Kelly spent over a year in space while his identical twin stayed on Earth, in a study into the physical effects of long-duration space missions.
NASA

Astronaut Sunita 'Suni' Williams exercising on the treadmill onboard the International Space Station during Expedition 14, April 2007. She was running the Boston Marathon while in space.
NASA

A prototype solar sail: LightSail 2,
built by the Planetary Society.
Justin Foley/Cal Poly/The Planetary Society

Artist's impression of a preliminary asteroid mining concept.
Yong Bin Tan

Remarkable new details of Jupiter have been revealed
by the Juno probe which arrived in 2016.

An artist's impression of a lander on Jupiter's moon Europa sometime in the 2030s.

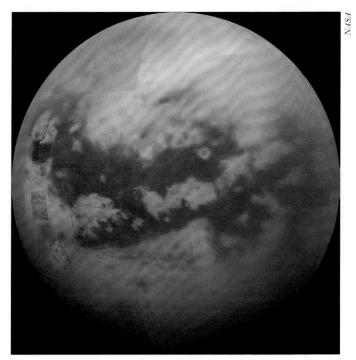

Saturn's major moon, Titan. Image taken in infra-red light from
the Cassini spacecraft on 13 November 2015. The infra-red light
penetrates the clouds and allows the surface to be seen.

NASA's Dragonfly mission will fly through Titan's skies in 2034.

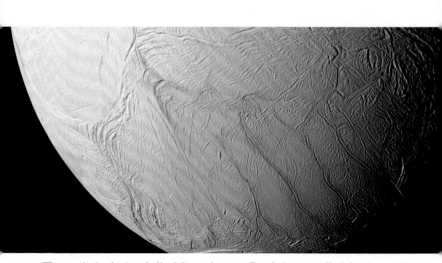

The cracks in the icy shell of Saturn's moon Enceladus are called the tiger stripes.
They are the source of plumes of water crystals from an under-ice ocean.
NASA

In March 2008 the Cassini spacecraft flew through
the plumes of Enceladus, making measurements.
NASA

a low point in the study of the red planet. But that was to change. By sheer chance Mariners 4, 6 and 7 had missed the planet's most spectacular features. Mars had wonders waiting.

Few people had their eyes on Mars during the later years of the 1960s. This was the time of the Moonshots. There were difficult decisions to make. In May 1971 – the month Mariner 9 was launched – NASA's budget was frozen. The following year Nixon authorised NASA to start working on the space shuttle. It ate into NASA's budget. Planetary missions already approved, such as the Pioneers to Jupiter and Saturn and the Vikings to Mars, were safe but the money dried up for missions after Voyagers 1 and 2 were launched to the outer planets in 1977.

Mariner 9 was dispatched to Mars in the hope that it would not repeat the findings of the previous Mariners. It turned on its cameras on 10 November when 800,000 kilometres from Mars. To the surprise and dismay of many it could hardly see anything except the faint outline of the southern polar cap and another faint dark spot. It had arrived during a global dust storm that had started in the southern hemisphere in September. Soon this merged with another growing storm coming from the northern hemisphere. There was no choice but to switch off the camera to save power and wait.

When the James Caird II passed over the Hellas basin the crew would have known that down there on its western rim was the wreckage of the first probe to touch Mars.

Mars 2 should have been the most spectacular probe to land on any planet up to that time. It had a spherical landing capsule and a heat shield. It had a TV camera with 360° view, as well as instruments to study the composition of the soil, the temperature, the atmospheric pressure and the wind. It had a mechanical scoop and a pennant with the state emblem of the Soviet Union, and even a small rover connected by a 15-metre umbilical.

On 27 November 1971 the descent module separated from the orbiter and headed for the Martian southern uplands. If all went well it would reach the surface in four-and-a-half hours. But the descent angle was too steep and then the parachute did not deploy. The first man-made object to touch Mars crashed in a region known for its dust storms and many chasms. By now it must be buried in dust; even so, people have looked for it. Its mothership remains in orbit to this day.

A few days later came its sister ship Mars 3. It's wise to send two at a time to Mars. This time the landing worked. It began transmitting after 90 seconds but stopped after 20 seconds, sending back a single 70-line image with no detail. Then nothing. It might have been spotted in Ptolemaeus crater on the other side of the planet from its sister, the parachute, retro-rocket, heat shield and the lander itself strewn over a landscape of ice and dust.

Mariner 9 waited for the dust to settle. Slowly Mars unveiled what we now call the Valles Marineris – the Valley of the Mariners. It was completely different from the images sent back by the earlier spacecraft. Tantalising at first – a long whitish streak was seen east–west near the equator. The dust continued to settle, and it was realised that the whitish streak was airborne dust filling a gigantic canyon system, a huge fault in the Martian crust that made the Grand Canyon on Earth look tiny. It took many weeks for the dust to clear. The streak was revealed to be a set of linked canyons more than 1,000 kilometres long and two kilometres deep. It was named after the spacecraft that discovered it. By the time the dust subsided in 1972, large parts of the planet's northern hemisphere had been revealed as plains much more sparsely cratered than those over which the first three Mariners had passed. Features known from Earthly observation, like bright Argyre and Hellas, turned out to be impact craters. As the pictures came into NASA's Jet Propulsion Laboratory in California, Carl Sagan took a Polaroid of the computer screen and rushed to the geologists' room.

Narrow valleys ran for hundreds of kilometres across the plains, sometimes on their own, sometimes feeding into networks of smaller valleys suggestive of drainage patterns. There were canyons that looked like cataclysmic torrents had torn them apart. It became increasing clear that water, and often wind, had played a major role in shaping the Martian landscape. There were streaks where dust had covered the surface and deserts with sand dunes. In some places there were full-blown dune fields. And the clouds of carbon dioxide billowing off Olympus Mons in the morning light – no wonder Schiaparelli thought of snow on a great mountain. Then there were the polar caps expanding and contracting in counterpoint with each other and the seasons.

The next thing to do was to land.

The landers were called Viking 1 and 2 and both faced a fraught landing site selection. For Viking 1, the original site seemed unacceptably risky. Eventually geologists found a suitable spot, still in Chryse Planitia somewhat further from the confluence of the four ancient channels. The delay prevented a landing on 4 July 1976, but it was generally agreed that a crash landing on that date would have been an unsatisfactory 200th birthday present for the United States.

Viking 1 landed successfully, and scientists eagerly awaited its first picture, which was to be of one of its footpads – in case it were to sink into Martian quicksand. The next picture showed the landscape and it didn't disappoint. It was stark, lovely and red! It was strewn with boulders thrown out in the creation of a crater somewhere over the horizon; there were rocks that obviously had been repeatedly covered and uncovered by windblown dust. The horizon of Mars could be seen, with low, arid hills off in the distance. Magnificent.

Just under two months later, on 3 September 1976, Viking 2 landed on Utopia Planitia, 6,760 kilometres away to the north-east on the other side of Mars. Overhead, the Viking orbiters took pictures. It was a Martian renaissance.

But one by one, inevitably the machines died. The Viking 2 orbiter suffered a propellant leak and was deactivated just twelve days before its second anniversary. The Viking 2 lander suffered a power failure after three-and-a-half years. The Viking 1 orbiter lasted four years and two months before running out of fuel. But the Viking 1 lander should not have died when it did. After almost six-and-a-half years of operations there was an error in a batch of code sent to it – an errant command that caused its radio dish to rotate down towards the sand, whereupon contact was lost. It would have continued to stare downwards until it ran out of power.

The Viking landers found no life – some strange chemistry, but no biological activity, and this disappointed many. Some turned away from Mars, avoiding it like an awkward personal encounter with someone who had disappointed them.

FASTER, BETTER, CHEAPER

———•———

I have already spoken of the 'Mars Underground' move-ment to keep the interest in the red planet alive in the 1980s. One of the dreams was a mission called the Mars Geoscience/Climatology Orbiter, which eventually became Mars Observer. One of its ambitions was to obtain a detailed picture of the planet's surface using radar. It was a struggle to get funded but eventually it was set for launch in 1990. However, after the Challenger disaster and the disruption it caused, it was decided to wait until the next time the planets were correctly aligned, two years later.

When an expensive spacecraft is delayed it is not just a question of putting it in a box and waiting to take it out again. Delay means the spacecraft's costs increase, and with no extra money cuts have to be made. It was not possible to disband the teams already at work on Mars Observer. To make savings the two heaviest instruments were removed, one of which was the radar. The craft was finally launched in

1992 with a laser ranging instrument called MOLA. But after surviving the coast to Mars, eleven months later, having been told to pressurise its fuel tanks to go into orbit, it fell silent, never to be heard from again. It probably exploded. I recall the loss of Mars Observer. It was a devastating blow. But the head of NASA Dan Goldin didn't see it that way.

Goldin was fed up of a space program that launched only a couple of super-expensive planetary spacecraft every decade. He wanted simpler probes so that if they failed it was a smaller loss than any of the giant probes they had been using. He came up with the 'faster, better, cheaper' philosophy. Smaller probes could, he maintained, expand space exploration – he believed that this way more could be done with the same amount of money. It was not, on the whole, a good idea. The first faster-better-cheaper program was called Discovery and the second Discovery mission was a Mars lander and rover – Mars Pathfinder – due for launch in 1997. It was to be followed by the faster-better-cheaper Mars Surveyor program.

Pathfinder, protected by airbags, bounced to a halt on the rocky Ares Vallis outwash plain. After an hour and a half, the lander unfolded. It was still dark, as the Pathfinder landed early in the Martian morning – about 3 a.m. local time. Not much more could be done until dawn. It was 4 July 1997 and it caused a sensation on the fledgling internet. The exact landing spot was where the ancient Ares Vallis channel opened up onto Chryse Planitia – the Plain of Gold where Viking 1 had landed 845 kilometres away.

As the Sun rose over Ares Vallis it soon became warm enough for Pathfinder to begin sending back the first images, which took twenty minutes to get to Earth. On Pathfinder's second day on Mars it was time to deploy the Sojourner rover. During the day it reached the end of the ramp and stopped, spending the night there. The next day, after a few wheel tests, the rover moved onto Martian soil and headed for a rock that had been dubbed Barnacle Bill some fifteen inches away. It was clear that the landing site had once been flooded: water deposits were seen, and the nearby rocks were of many different types as if transported there from many regions of Mars.

After an extraordinary performance, by early October Pathfinder was in trouble: its battery was failing, affecting most of its systems. A week later it became clear its days were numbered. In its three months it showed that its landing site had been awash with water 3–4.5 billion years ago and dry for the last billion years. It sent back more than 17,000 images from the lander, and the rover performed sixteen detailed examinations of rocks.

The next mission to Mars was called the Mars Global Surveyor and it carried copies of five of the lost Mars Observer's instruments, for the most part made of spare parts. Soon after arriving at Mars it carried out a new method of entering orbit: it started a long series of passes through Mars's thin upper atmosphere. It was a way of losing energy, making its orbit circular. 'Aerobraking', as it is called, was risky. Before the arrival of faster-better-cheaper thinking, rocket engines were used, but atmospheric drag costs nothing. It took a lot longer than anticipated. The drag was produced by the craft's solar panels and the arm holding one of them was faulty. It was not until early 1999 that MGS reached its final orbit of two hours and 400 kilometres above the surface.

Soon the infra-red spectrometer looked for minerals, the camera looked for signs of water and the laser zapped the surface ten times a second. In three months, its laser produced 27 million altitude measurements, revealing Mars's topography like never before.

The poles were very different from each other. The surface at the north pole is probably less than 5 million years old. It is riven with canyons and valleys gouged out by winds and by evaporating subsurface ice causing collapse. But there isn't that much water ice at the north pole. The south pole terrain is probably much older and on average about 6 kilometres higher. The south pole has been mostly sculpted by wind.

Excited by the bounty from Mars Global Surveyor in March 2000, planetary scientists gathered in Houston for the annual Lunar and Planetary Science Conference. At least a hundred papers on Mars were presented, most of them based on MGS data. Mars was changing before their eyes. The idea that there had once been an almost global ocean on Mars was taken seriously and likewise the prospect that, far from being geologically inert for billions of years, Mars could still be active. Mars Global Surveyor was a great success, but the program it had spearheaded was a disaster.

The faster-better-cheaper spacecraft Mars Climate Orbiter and Mars Polar Lander were both lost due to stupid mistakes. The Mars Climate Orbiter was launched in December 1998. When it went to aerobrake into orbit it went in too close and broke up. The reason why it passed too close to Mars was because of a miscommunication between two sets of engineers. One team was using metric units, the other imperial units and someone fluffed the conversion. Mars Polar Lander was launched in January of 1999 and lost in September 1999. It seems that a kilometre or so above Mars's polar plains, the spacecraft was descending, using its rocket motor to reduce speed. As expected, its landing struts deployed. However, it seems the rocket and the landing gear had been tested separately, but not together. The struts caused the rocket motor to switch off and the Mars Polar Lander crashed. Faster-better-cheaper was dead.

The implications were grave. It caused the United States to abandon its ambitious plans to bring back rocks from the surface of Mars before the end of the 2000s. The devastating news was given by Dr Carl Pilcher, the scientist leading NASA's planetary exploration program.

'The search for life on the red planet will have to slow down until people on Earth have worked out how to land on Mars without crashing,' he said, and described the timetable which called for a sample return mission that would bring back the first Mars rocks to Earth in October 2008 as wildly optimistic: 'The jury is out on whether we have the technological capability.' Breaking the bad news to space scientists at a conference in Houston, Texas, Dr Pilcher had on a sweatshirt that declared: 'Obey Gravity: It's the Law.' He would not discuss the results of the study into the loss of the Mars Polar Lander, due to be published at the end of the month, but he said the report's conclusions 'make sober reading'. 'Our next lander on Mars had better work,' he added.

Despite the bad news, the identification by the Mars Global Surveyor of hematite – an iron mineral formed in the presence of water – was a landmark. Along with the many indications of past running water on Mars, it suggested there was an ancient era of oceans. The possibility of it having once harboured life had increased.

By now it was felt by many members of the European Space Agency that it was about time that they went to Mars and they came up with the Mars Express mission. The name referred to the rapidity with which it had to be built. This was because there was an opportunity approaching for a swift transfer to Mars as both planets were aligned in a convenient way. The design and instrumentation owed a lot to a failed mission of 1996. The Russian Mars 96 mission failed to reach orbit and since most of that mission was provided by European scientists it was relatively straightforward to transfer the elements to another spacecraft. Because of this the cost of the mission was $345 million, probably half the cost it otherwise would have been.

The Mars Express craft arrived at Mars in 2003 and a team of mostly Italian scientists used its radar to infer that there is a lake-sized reservoir of water with sediment beneath the southern polar cap. According to one of the team, Elena Pettinelli, a researcher at Roma Tre University in Italy, 'This discovery is changing again the view of the possible presence of liquid water.' During polar passes between May 2012 and December 2015, the European radar saw 'anomalously bright subsurface reflections'. Other scientists, especially those working with data from the radar aboard NASA's Mars

Reconnaissance Orbiter (of which more later), were unconvinced. Theoretically, the American spacecraft's radar should also be able to detect this underground lake if it exists, but it has not. The Mars Express radar instrument can penetrate the top 100 metres below the surface while the instrument on the NASA spacecraft sees down to 10 metres below the surface, so that might be an explanation for the discrepancy.

Mars Express has been a great success with its high-resolution images, its mineralogy mapping and radar sounding of permafrost regions. It's still going today. But it took a hitchhiker to Mars which was not so successful. On its side, to be deployed at Mars, was the Beagle 2 lander.

What can one say about the Beagle 2 mission except that it was like none other? It was the idea of Colin Pillinger, a professor at the UK's Open University who was a specialist in analysing samples from the Moon. Beagle 2 was to be a small lander designed to look for life. It was an interesting proposal, but it couldn't get official support, so Pillinger decided to go it alone and raise the funds himself. He wanted companies, people – well, anybody – to sponsor the mission. Work started on the promise of money forthcoming but little ever did and eventually the UK government had to find £45 million to save the project, which thanks to Pillinger maintained a high public profile.

Beagle deployed successfully from Mars Express and headed for the surface but was never heard from again. There was an enquiry, which didn't reach any particular conclusion. In 2015 it was found in images taken from orbit. It appears to have landed successfully but didn't deploy properly. It almost worked. This was sad because the mission was run in such an outrageous manner by Pillinger it would have been justification had it worked. In my diary at the time I wrote, 'Some space missions have to be very unlucky to fail, this one will have to be very lucky to succeed.'

In 2003 the US dispatched two rovers to Mars. The first of them, Spirit, landed in the crater Gusev, not far from Elysium Planitia, in January 2004, on what was believed to be the bed of a former lake. As the airbag protection deflated Spirit took images of its position, among them the highest resolution image ever taken of another planet up to that time. The red surface was flat and littered with small rocks; the horizon was three kilometres away. In one direction there were a few low hills seen and they were named after the lost crew of the space shuttle Columbia.

Three days after landing, Spirit went up to its first rock, named Adirondack after a range in New York, and spent three hours drilling a hole 2.85mm deep. Analysis of the rock by several of Spirit's instruments indicated that the hard, crystalline rock was very similar to terrestrial volcanic basalt. Part of the cable shield for the drill was made of aluminium recovered from the World Trade Center after the 2001 attacks. The same is true for Spirit's sister rover, Opportunity.

During its wanderings of 7.73 kilometres, Spirit found evidence of only slight weathering on the interior plains of Gusev, and curiously no evidence that a lake had been there. The nearby Columbia Hills did show evidence of weathering by water and Spirit found sulphates and minerals – goethite and carbonates – which only form in the presence of water.

It is believed that Gusev crater may have held a lake long ago, but it has since been covered by igneous materials. All the dust contains a magnetic component which was identified as magnetite with some titanium. Everything had a thin coating of dust – dust covers everything on Mars and is the same in all parts of the planet. In late 2009 Spirit became stuck in a 'sand trap' at an angle that prevented the recharging of its batteries. Its last communication was on 22 March 2010.

Three weeks after Spirit landed, Opportunity set down on the other side of Mars, in Meridiani Planum, a vast, flat expanse just south of the Martian equator, poised between the red planet's southern uplands and its vast northern plains. It's geologically complex and strewn with minerals that usually form in shallow seas. Billions of years ago this region was flooded with water and several times it evaporated, only for the acidic, mineral-rich water to return to lay down thick layers of sediments. Eventually the cycle stopped, and it has been dry on the planum for the last billion years. It's a popular place to explore – even Mark Watney in *The Martian* passed through it. Currently, the region is restless as it is facing the Sun when Mars is at its closest to the Sun in its orbit. These are dust storm conditions, and once every few years a local dust storm can grow into a planet-encircling shroud obscuring everything on the surface.

Opportunity stayed active on Mars for much longer than Spirit: from 2004 to 2018, it traversed 73 kilometres, a record for a journey on another planet. It fortuitously landed in a small impact crater on an otherwise wide and flat plain. Its first goal was Endurance crater, then the larger Victoria crater, ending up at the even larger Endeavour crater in August 2011. It made many scientific discoveries, including finding a meteorite on the surface, but had a few difficulties along the way such as the time it got stuck in the sand in 2005, when several of its six wheels were covered in sand. It took six weeks of delicate manoeuvring to extract it. Early in June 2018 it was overpowered by a global dust storm which piled sand on its solar panels, depriving it of energy. It never recovered.

The Mars Reconnaissance Orbiter is a kind of a red planet analogue to the Lunar Reconnaissance Orbiter. It arrived at Mars in 2006, and it did for Mars what its counterpart did for the Moon. It saw avalanches in progress, dust devils moving across dusty plains, sand dunes that move, fresh impact craters and brine flows on steep slopes.

One of its most publicised, as opposed to scientifically valuable, images was of a region of Cydonia which is not far from Chryse Planitia. The area was first imaged by Viking 1 in 1976. Looking at the Viking image, someone noticed that an isolated mountain looked rather like a face. At the time scientists dismissed it, rightly, as a trick of light and shadow, but the 'Face on Mars' concept was born: a mountain sculpted by now-long-dead Martians as a temple or sign to the cosmos that they were there. Many people were therefore excited to see if new, higher-resolution photography from the Mars Reconnaissance Orbiter would confirm the idea. It didn't. Obviously, it's an ordinary hill. Our minds saw more than there was – as, incidentally, is the case with many mountains on Earth that resemble human features, such as Mount Mansfield in Vermont, New England in the USA.

MRO's radar measurements of the north polar ice cap determined that the volume of water ice there was 821,000 cubic kilometres, equal to about 30 per cent of the Greenland ice sheet. It also established that the impacts which caused some of Mars's newer craters had excavated relatively pure water ice. After exposure, the ice fades as it sublimates away. Ice was found in five locations, three of them in the so-called Cebrenia quadrangle – a region of the north-east portion of Mars's eastern hemisphere, occupying only 3 per cent of the planet's surface. Together the Mars Global Surveyor, Mars Odyssey and Mars Reconnaissance Orbiter have filled in the geological details of a complex planet covered in minerals laid down in a vast ancient ocean. The evidence is overwhelming, the chemistry unambiguous.

There are widespread deposits of chloride minerals formed from the evaporation of mineral-enriched waters, suggesting that lakes may have been scattered over large areas of the Martian surface. In its early days Mars could have been described as a waterworld of active volcanoes, rivers, rain and lakes. Usually chlorides are the last minerals to come out of solution as a lake evaporates. Carbonates, sulphates and silica should precipitate out first and in fact, sulphates and silica have been found by the Mars rovers on the surface. In 2009, a group of scientists reported seeing nine to ten different classes of minerals formed in the presence of water. Different types of clays which are formed in water have been found in many locations. Areas around Valles Marineris were found to contain hydrated silica and hydrated sulphates, again evidence of past water. Other researchers have identified hydrated sulphates and iron minerals in Terra Meridiani and in Valles Marineris. Other minerals found on Mars are jarosite, alunite, opal, and gypsum. In August 2011 NASA announced that MRO had detected what appeared to be flowing salty water on the surface or subsurface of Mars.

In 2010, MRO detected evidence of haematite within Gale crater and a few years later the Curiosity rover found an entire layer of haematite at the base of Aeolis Mons – the crater's central mountain, more often called Mount Sharp. Haematite is evidence that water once filled Gale crater, and there is evidence of water from other observations too. From orbit, layers of clay have been detected, and there are channels where water once flowed along the foothills of Mount Sharp.

Water was everywhere on Mars – but where has it gone? The answer will show how similar exploiting ice on Mars will be to exploiting ice on the Moon. What we learn on the Moon we can apply to Mars.

In late May 2008, on a vast, almost featureless, frozen plain high in Mars's northern arctic region, NASA's Phoenix spacecraft flexed its robot arm. Phoenix had been launched in August 2007 and touched down in a region where it was thought that water ice could be found just below the surface. The camera on the end of the arm looked underneath the craft, showing patches of white uncovered by the thruster's exhaust when the craft landed. Later on, it scraped at the coarse sand, exposing the white hard layer just beneath the surface. A few days later it looked back at the small trench. The whitish patches had almost gone, vaporised. It could not have been frozen carbon dioxide because given the conditions around the lander that would have disappeared very quickly. Could it be water ice? Frozen water just below the surface? Mars, it seems, has ice that in some regions is easier to get at than ice on the Moon.

Unlike other places visited by Viking and Pathfinder, almost all the rocks near Phoenix were small. The land was flat as far as could be seen; there were no ripples or sand dunes. Instead the surface was shaped into polygons 2–3 metres in diameter, bounded by troughs 20–30 centimetres wide, typical for a frozen surface dominated by ice expanding and contracting due to changes in temperature. The onboard microscope showed that the soil was made of rounded clay-like particles. Dust devils were seen scuttling over the surface in the mid-distance. Snow fell from cirrus clouds made of water ice crystals. Wind speeds ranged from 11 to 58 kilometres per hour. The highest temperature was −19.6°C, the coldest was −97.7°C. Phoenix was in a place on Mars never experienced before.

The surface soil was moderately alkaline, with modest salinity. A sample of the white material was scooped up and placed into a container to be heated. When the sample exceeded 0°C water vapour was detected. Chlorides, bicarbonate, magnesium, sodium, potassium, calcium, and sulphate were detected during the soil analysis. A significant finding was that while there was ice everywhere just below the surface, the soil had not interacted with liquid water of any form for as long as 600 million years. Phoenix was sitting on a plain that had been frozen and dry for over half a billion years. Tests showed nothing lived in the soil.

A big surprise was that perchlorates were found in the soil. Perchlorate is a compound that is toxic to humans and breaks down anything organic. It might be nasty to life but it could be used as the basis for rocket fuel. Mixed with water, perchlorate can lower the freezing point of water, like salt does when applied to roads to melt ice. It has been speculated that gullies, which are common in certain areas of Mars, may have formed from perchlorate melting ice and causing water to flow down on steep slopes. When perchlorate burns it produces a molecule called chlorobenzene. In 2013, NASA's Curiosity rover discovered chlorobenzene in Gale crater and a retrospective analysis of the Viking 1 soil analysis suggests it was also found in 1976.

Phoenix was not designed to survive the Martian winter. Its landing zone is in an area that is usually part of the north polar ice cap during the winter, and later the lander was seen from orbit encased in dry ice, its fragile solar panels likely broken off under the weight of the ice.

It appears that just a few centimetres below the surface at latitudes as low as 35°N is an abundance of ice – about a third of the Martian surface contains shallow ground ice. Some of these deposits are up to several hundred metres thick and many appear to consist of nearly pure water ice. The water that had once so dominated Mars and left so many signs of its action was still there. Mars's lost ocean had been found, frozen underground.

But what of life? If there is life on Mars, then it is surviving against the odds. A major problem for it is that Mars lacks nitrogen, which is essential to produce proteins, without which life as we know it cannot develop. The Curiosity rover did find nitric oxide in the sediments of Gale crater, but it seems to have got there as a result of non-biological processes such as during meteor impacts or lightning strikes in the distant past, when Mars had a much wetter and thicker atmosphere.

Considering the evolution of Mars and the Earth, they will have been very similar when they were young, temperate and wet. Some scientists consider the early Mars to be a better candidate for life than the Earth. But the two planets evolved in different ways. Mars is smaller than the Earth and its molten core – where the planet's magnetic field is generated – froze very quickly. On our planet it didn't and because of the insulating effect of our thick mantle, is not that much cooler than when it formed.

Earth has a vigorous dynamo that generates a magnetosphere that shields the Earth from solar radiation and the solar wind. Mars lost that magnetic shield very early on and its atmosphere was ablated by the solar wind and is now a mere 1 per cent of that of the Earth. The lack of a shield allows harmful radiation to reach its surface, sterilising the top metre or so. Only away from the surface, underground or possibly in caves, could lowly forms of life cling on, the final survivors of a once-living world with no future.

When our worlds were young our solar system contained far more debris than it does now, and large collisions were common. Many worlds were destroyed, none escaped intact. If life developed on Mars or Earth it is possible that it could have been taken to the other planet by an interchange of rocks between us. We have identified meteorites on Earth that were blasted off the surface of Mars and conversely there must be Earth meteorites on Mars. Perhaps if we find evidence for present or past life on Mars we may be looking at life that began on Earth, or we could be looking at our ancestors before they came here, making our Mars mission a journey home.

CLIMBING
MOUNT SHARP

———•———

Intrigued by the geological signs seen from orbit, NASA landed its next rover, called Curiosity, in Gale crater in August 2012. Curiosity is the most sophisticated rover yet sent to the red planet, with a range of instruments to look at its weather and geology. At the time of writing it has survived over 2,786 sols, and travelled over 20 kilometres. You can see the wear and tear on its metal wheels which mark out a unique repeating pattern in the Martian soil – Morse Code for JPL, the Jet Propulsion Laboratory in California where it was made.

Curiosity is climbing a mountain believed by geologists to be an enormous mound of sedimentary rock layers deposited over 2 billion years. At 5.5 kilometres high this mountain is about the same height as the tallest mountain on the Moon – Mons Huygens, near where Apollo 15 landed. From rim to rim, Gale is 154 kilometres across and is an ancient – 3.5–3.8-billion-year-old – impact crater. It's thought it was filled with sediments, first by water and then by wind. Then wind erosion scoured out the sediments, leaving Mount Sharp isolated in its centre. It's hoped that during Curiosity's trek it will be able to study 2 billion years of Martian history in the sediments.

Additionally, the rover's landing site was an alluvial fan, created by water. It was a region unlike anything seen by previous Mars landers, looking like a dry lakebed. Indeed, the crater's floor is covered with deltas and fans carved by ancient water. That is where Curiosity first drilled, making two small holes in the Martian soil. When the sample was analysed it was found that the rock contained clay minerals, a clear sign of water.

In September 2014, Curiosity reached the base of Mount Sharp, its main destination. Watching from above in 2015, scientists spotted what they call recurrent slope lineae, downslope flows of briny water, on Mount Sharp near Curiosity. The rover is now halfway through a region called the 'clay-bearing unit'. In the six years since it started to ascend Mount Sharp, it has been going higher and higher into the sediments, finding ever-muddier layers. Geologists want to know how high these mudstone layers go. When Curiosity gets through the clay layer it's thought it will reach a sulphur-bearing layer. It might not get to the top of Mount Sharp because in a few more years its ever-declining nuclear power system will degrade enough to limit operations. Eventually Curiosity will get stuck, almost certainly never again to be touched by human hands.

After a flawless launch and a quiet half-year cruise to Mars, the InSight spacecraft landed safely in Elysium

Planitia on 26 November 2018, carrying a seismometer to detect Marsquakes as well as probes to investigate the deep interior of the planet. It was another mission that reused parts of previous craft, in this case those that had been used on the Phoenix lander. InSight found that Mars is alive with quakes, trembling more often than expected. It has so far detected more than 450 seismic signals. Because Mars's crustal structure differs from that of the Earth – it doesn't have tectonic plates – the quakes probably originate in volcanically active regions. A pair of quakes has been linked with the volcanic region Cerberus, where boulders have been seen at the base of cliffs having been shaken loose higher up.

But the news from InSight is not all positive. On 28 February 2019, one of its instruments, called the Heat Flow and Physical Properties Package, started drilling into the surface, intending to reach a maximum depth of 5 metres after two months work. After only a week or so it came to a halt having reached 35 centimetres. After months of studying the problem it was concluded that the soil on Mars does not provide the necessary friction for drilling, causing the mole to bounce around and form a wide pit around itself rather than digging deeper. They tried to increase the friction and therefore its digging ability by pressing it against the side of the hole, but the mole backed out of its hole after a few weeks.

The year 2020 will see three spacecraft dispatched to Mars, all heading for the surface, for the next phase of Martian exploration has begun. We have flown past it, observed it from orbit, landed on it, roved around it, poked, prodded and drilled it and now it is time to prepare the way to bring a piece of it back to Earth.

Inevitably there are limitations to the type of science experiments that can be put on a lander or a rover. Even so, scientists have proved themselves ingenious at cramming in the most sophisticated tools. We have sent cameras capable of multi-spectral views; we have fired lasers at rocks and analysed the light given off to provide information about the rock's composition; we have built instruments to monitor the weather conditions: humidity, pressure, temperature, windspeed and ultraviolet light from the Sun; we have used devices that irradiate samples with alpha particles and analysed the X-rays given off by the collision, which provide a clue to the composition of the samples. We have gathered samples and poured them into a chamber in which they are irradiated with X-rays to determine mineral structure. We have sent instruments that look for organic molecules, instruments that monitor radiation, instruments that look for ice. But all the scientists involved in this, having spent years miniaturising their instruments and detectors, getting them to work on a very limited power budget, in very little space, let alone making them tough enough to stand the rigours of a rocket launch and interplanetary flight – all of them would rather get a piece of Mars in the laboratory on Earth to analyse. That time is coming.

Mars 2020 is a NASA Mars rover mission with a planned launch on 17 July 2020. It will touch down in Jezero crater on Mars on 18 February 2021. Jezero is on the edge of the Isidis basin and is again thought to have been once flooded with water. Like Gale crater it has a delta and there are shapes on the surface that look like the patterns you get when mud dries out. It's an old crater – perhaps life developed there. The rover for this mission, Perseverance, is based on Curiosity but with improved instruments and a few remarkable additions. It will carry a helicopter drone called the Mars Helicopter Scout. It will reconnoitre the local area, providing information as to the best route for the rovers, both Perseverance and another one that will be landed as close as possible to its route sometime in the near future. It is expected to fly about five times during its 30-day test period. Each flight will take no more than three minutes at altitudes of three to ten metres. If it works, helicopter drones will be a part of future Mars missions because they can move quicker and see farther than any rover.

A key task for Perseverance will be gathering samples to be picked up and taken to Earth by a future mission. Between 20 and 30 drilled samples will be collected and cached inside small tubes for later retrieval by NASA in collaboration with the European Space Agency. A 'fetch rover' would retrieve the sample caches and deliver them to a Mars ascent vehicle (MAV). The MAV would launch from Mars and enter a 500-kilometre orbit and rendezvous with a new Mars orbiter. The sample container would be transferred to an Earth entry vehicle (EEV) which would bring it to Earth for analysis.

Perseverance will also test a technology that will become vital for the human colonisation of Mars – a device to produce a small amount of pure oxygen from atmospheric carbon dioxide. When humans get to Mars, they will have to live off the land so making oxygen this way will be essential. It's called MOXIE (Mars OXygen In situ resource utilization Experiment), and the version on Mars 2020 is said to be a 1 per cent scale model. It uses a process called solid oxide electrolysis. Mars has a 96 per cent carbon dioxide atmosphere and MOXIE will attempt to produce 22g of oxygen per hour with greater than 99 per cent purity for 50 Martian days. If it works, one can envisage a version 100 times larger powered by a radioisotope generator. It could be placed at a Martian landing site before the crew arrives, to produce up to 2kg of oxygen an hour, which would be stored for the crew to use when they arrive. It could be used for life support and as a component of rocket fuel. For the long-term exploration of Mars, never has a more important instrument been placed on its surface. Carbon monoxide is also produced as a by-product. This could be used as a propellant, either as carbon monoxide or being converted to methane first.

The European Space Agency was to have launched its ExoMars mission containing the Rosalind Franklin rover in 2020 but has now delayed until the next launch window in 2022. The delay has been brought about by the parachutes. A landing on Mars is tricky because it's a planet with an awkward combination of gravity and atmosphere. If you want to enter the Earth's atmosphere from orbit, or from an incoming trajectory from the Moon or Mars, you have to slow down, and initially this is done using a heat shield, which glows red from the heat of re-entry. After that has been done the rest of the descent is by parachute, or in the case of the retired space shuttle, by gliding. You can't do that on Mars.

For a start, Mars's atmosphere, despite being much thinner than the Earth's, reaches much further into space because the pull of gravity is weaker. This means that a heat shield would only partly work – it would only slow you down a little. Once the heat shield has done its work and is jettisoned you would still be travelling too fast and still need to lose speed. There are various ways this can be done. One is by parachute, but the air is still too thin to use a parachute all the way down, even if you are aiming for the regions of Mars where the terrain is lowest and the atmospheric pressure the greatest. You need something else. Mars Pathfinder, Spirit and Opportunity did it with airbags. Curiosity did it with thrusters attached to a 'sky crane' that lowered it to the surface.

The European Space Agency has an unfortunate experience of landing on Mars. To test the Martian entry system for the Rosalind Franklin rover they built a smaller Mars probe called Schiaparelli and sent it to Mars, hitching a ride with the ExoMars Trace Gas Orbiter spacecraft in 2016. In October of that year it was due to touch down in the Meridiani Planum region to test its entry, descent and landing system.

If you were standing on the reddish sand of Meridiani Planum on a particular afternoon that year you might just have caught a dark speck in the distance moving slowly against the peach-coloured sky. It's Schiaparelli making its six-minute descent through the atmosphere. Its heat shield would be glowing, for just minutes earlier it was hurtling through the vacuum of space at 21,000 kilometres per hour. But then something went wrong and the lander's guidance failed (it thought the probe was below the ground) and it prematurely jettisoned the back heat shield and the parachutes. The braking thrusters that were to fire close to the ground for 30 seconds fired for only three seconds. Then things got worse. While still 3.7 kilometres above the ground Schiaparelli thought it had landed and switched on its ground systems. Then all contact was lost, and the probe crashed into Meridiani Planum at 540 kilometres per hour. Later MRO saw the crash site. Although the lander crashed, officials at the European Space Agency declared it a success because its objective of testing the landing system was fulfilled! A few years later, in August 2019, the parachutes failed again as they tore during a drop test in Sweden. The problem had been seen on a previous test and modifications made to the parachutes, but it seems they didn't solve the problem.

Whenever it is ready to launch, the ExoMars mission containing the Rosalind Franklin rover will look for signs of past and present life. It's headed for a place called Oxia Planum, a 200-kilometre-wide clay-bearing plain located on the south-west margin of Arabia Terra. The centre of the landing ellipse overlaps the Cogoon Vallis drainage system towards the northern plains.

Is there, in some ancient clay-rich deposit billions of years old, a tiny speck of life? A primitive organism's remains waiting to be discovered? The molecular machinery required to build such an organism would be very complex, and if for instance ExoMars did find life in 3.9-billion-year-old deposits, it would show how swiftly, in the right conditions, life can find a way to exist.

Then we could look to investigate the idea raised in the previous chapter, of an Earthly ancestor having seeded life on Mars – or the reverse – through an exchange of rocks. This, though, is something no planned rover will be able to ascertain, instead requiring a sample return mission and DNA analysis back on Earth. Even then, we may not obtain any conclusive results. But whatever unanswered questions we may be left with, to know there has been life on another world would be one of the greatest discoveries.

THE GULF OF SPACE

L et us imagine that the James Caird II begins its 201-day return to Earth in 2040. The crew are exhilarated by their encounter with Mars but back on Earth mission control have their worries.

Whatever the design of the Mars Transit Vehicle, it will take advantage of the incredible advances that will be made in artificial intelligence over the next two decades. Built into the spacecraft's AI control system will be a crew evaluation package, designed to look at every crew member's activity and every interaction with the computer. Concentration, reaction times, mood, leisure activities and the many biomedical tests crew members will carry out on each other will be evaluated to see how they are doing individually and collectively.

The Mars return voyage will be a question of balanced survival and timing. Resources and tensions will have to be managed to ensure no changes take place and the status quo is maintained for the best part of a year. It is likely that the crew's evaluation scores will decline at the start of the voyage home given the lengthy journey to come.

The scores of the James Caird II crew have been declining since before they reached Mars and have continued that way even with the relief of the start of the homeward leg of the mission. The decline was expected – part of the reason why the first mission was a Mars orbital mission was to study this effect. It was clear that without the information from a flyby mission, the stress and many complicated tasks involved in landing, living on Mars and returning to the mothership in orbit was a stretch too far for any crew. Concentration, attention span, mood, irritability and growing intolerance were always potential mission problems. Flight controllers were glad the crew were to have light duties throughout their return, but there were still the critical manoeuvres to be made at the end of the journey when the James Caird II had to be captured by Earth's gravity. Things would be OK, the mission planners told themselves, provided nothing major happens on the way home. But the story of exploration is that nature can fight back and lay waste to the best laid plans.

The best time to go to Mars is when the Sun is at minimum activity. It is less threatening then due to the relative magnetic calm on its surface. By 2040 we might expect the Sun to have been experiencing general low activity for at least 30 years, going through a prolonged minimum, as historical records suggest it does from time to time.

The James Caird II mission planners had been glad of the minimal solar activity. An interplanetary transit during solar maximum would have been out of the question without a new design of Mars Transit Vehicle – one considerably more complex and larger.

But the Sun does not always follow the behaviour we wish of it during solar minimum and it's conceivable that a 2040 return trip from Mars could coincide with unexpected and unwanted solar activity.

We can imagine a small region on the Sun's surface pulsing with hints of a new spot group. Those astronomers who study the Sun every day are swiftly onto it, focusing their ground-based and satellite telescopes, looking to see if it becomes a problem. Observed by Sun-orbiting satellites, the group rounds the limb as seen from Earth. McMath Region 13970 is bristling with magnetic intensity. There is fascination and apprehension. A crew of six is in between the planets.

A few days later the region has grown to become a rare blotch on the Sun. Our star is no respecter of statistics. Suddenly the region flares due to the collapse of magnetic tubes that have risen in an arcade above the Sun's surface. The magnetic energy is explosively converted into a stream of fast-moving particles, by chance headed for the vicinity of Mars. To paraphrase H.G. Wells, the chances of it hitting Mars were a million to one, they said. Some of the energy registers as white light flares, a form of extreme flaring first observed by Richard Carrington and Richard Hodgson in 1859 during what is now famously called the Carrington Event. Then, the region had the classic magnetic 'delta' spot configuration, and as well as the flare radiation, two coronal mass ejections or CMEs – or, as the press called them, 'plasma cannonballs' – were thrown into space. Their trajectory took them away from the Earth but, again, towards Mars. The fact is that despite observing the Sun for hundreds of years and in exquisite detail for decades, we do not know what it could throw at us.

Sunwatchers know very little about the big events. The largest event of modern times took place on 23 February 1956. The radiation did not affect society much as we were far less dependent on electronics and there were fewer passengers in aircraft flying at high altitudes than there are today. It was classified as a once-in-150-years event. Today we have electronics everywhere, they are getting smaller and operate on tiny amounts of electric charge and current. Over the years radiation from the Sun has frequently damaged spacecraft in orbit, sometimes leading to a total loss. In July 2012 another Carrington-style event occurred on the surface of the Sun, but this time nobody noticed as it was aimed away from the Earth. As its fury and its radiation blasted from the opposite side of the Sun, the opening ceremony of the Olympics was taking place in London.

The weakened crew of the James Caird II are put on high alert and storm protocol initiated. In addition to the radvests worn by all astronauts since the early 2020s they are going to need extra protection. The spacecraft is turned so that its densest part faces the onslaught. Everything that can be moved is placed in that part of the ship, but the estimated radiation levels are higher than have been planned for, higher than thought possible.

The coronal mass ejections arrive at the James Caird II in only 28 hours – a record. Travelling through interplanetary space, they clear the path for any subsequent ejecta.

In their wake they leave a tenuous trail of energised protons. Communications become erratic – lots of interference, but what data there is indicates the computers are taking a pounding. Charged particles are striking microchips through their radiation shielding causing what are called bitflips, latchups and burnouts. The solar panels are damaged, losing 15 per cent of their capacity within an hour. Readings from the dosimeters indicate the crew are going to get very sick. It is probable they will all develop cataracts.

Satellites around the Earth indicate that something strange is happening in interplanetary space. Galactic cosmic rays are not arriving as expected and observers realise that a very large and tenuous plasma structure capable of deflecting them is on the loose and is also accelerating ions from the wake of the previous CMEs. The accelerating particles are scattering between the converging solar wind structures, gaining energy with each bounce.

On the James Caird II it is only after the chaos of the first two shocks and magnetic impulses that the CMEs arrive. The record of their arrival is muddled. By that time some of the craft's radiation-monitoring systems are compromised or have failed outright. They are still 163 days from Earth.

THE WEAKEST LINK

———————•———————

Just how dangerous is a mission to Mars took us a while to realise. The signs had been there, even if good data was wanting, but there was optimism that technology would sort out any problems. We had sent many spacecraft to Mars which had monitored the conditions between the planets. Radiation didn't seem so bad and the voyage of about 200 days was shorter than many astronauts had spent in space aboard the International Space Station. But as spacecraft designs came and went and more astronauts carried out long-duration missions onboard the space station in its radiation-shielded orbit inside the magnetosphere, there grew the realisation of what was the greatest barrier to getting to Mars. Such a voyage exposes mercilessly the weakest point in the entire system – us.

It is a truism that we are not designed to live in space. 'Modern' *Homo sapiens* (that is, people like we are now) first walked the Earth about 100,000 years ago. Since then humanity has totalled 107 billion people, meaning that for every person alive there are fifteen that are dead. Of these

107 billion, fewer than 600 have been into space and only a small fraction of those have been on long-duration flights. During the 77 person-years humans have spent in space we have learned a lot about the human body's adaptation to this unnatural environment, but the more we know the more we realise we need to know what happens to us in far more detail if we are to go to Mars.

Some changes are rapid and occur immediately upon entering space, such as the shift of fluids to the head. Astronauts call it 'puffy face, bird legs'. The effect plateaus after a few weeks but others do not. There are important changes to the left ventricular region of the heart. One would expect heart changes as the nature of pumping blood around the body alters when gravity is absent. The thickness of the heart wall increases and, as far as we can tell, the effect is progressive.

Astronauts also suffer from a loss of bone density, weak-ened muscles and problems with nutrition, as well as changes in the immune system, affected sleep patterns and psycho-logical disturbances. Bone loss has been observed during spaceflight since at least as early as Gemini in the 1960s. Although most early measurements were not reliable, bone loss has now been detected in Gemini, Soyuz 9, Apollo, Skylab, Salyut 7, Mir, and the International Space Station missions. Astronauts lose an average of more than 1 per cent bone mass per month spent in space. There is concern that during a long-duration mission, too much bone loss, along with elevated serum calcium ion levels will result in irrevers-ible skeletal damage. William E. Thornton, an astronaut and physician, was one of the biggest proponents of exercise as a way of preventing bone loss. Thus, treadmills and resistance devices are used to limit muscle atrophy. But perhaps there is a limit to their effect.

Scott Kelly had been in space three times before the big one. His first flight was as pilot of space shuttle Discovery in December 1999, the third servicing mission to the Hubble Space Telescope. The second was in 2007 as commander of space shuttle Endeavour on a trip to the International Space Station, a place he would come to know well. His third was in a Soyuz capsule launched in October 2010. He assumed command of the ISS a month later and returned home in March of the following year. Then came the big one.

Scott came from a remarkable family. His identical twin brother Mark was also an astronaut, a veteran of four space shuttle flights from 2001 to 2011, spending 54 days in space. The idea was to compare the two while one was in space for a year and the other back on Earth, to see what changes long-duration spaceflight made to the human body. Along with cosmonaut Mikhail Kornienko, Scott took off in a Soyuz in March 2015 and returned to Earth a year later. After the mission, adapting back to Earth's gravity was not easy.

'Every part of my body hurts,' he wrote in his autobiography. 'All of my joints and all of my muscles are protesting the crushing pressure of gravity. I'm also nauseated, though I haven't thrown up. I strip off my clothes and get into bed, relishing the feeling of sheets, the light pressure of the blanket over me, the fluff of the pillow under my head ... I've been to space four times now ... I spent 159 days on the space station in 2010–11. I had a reaction to coming back from space that time, but it was nothing like this.'

The Astronaut Twin Study monitored Scott for 5,356 revolutions of the Earth. The results were released in 2019 and in general they are not encouraging for long-duration spaceflights. There were several long-lasting changes including changes in DNA expression and, worryingly, problems with cognition.

Scott Kelly and Kornienko kept a journal. Kornienko wrote, 'The thing you miss there most of all is the Earth itself, I missed smells. I missed trees, I even dreamt of them. I even hallucinated. I thought I smelled a real fire, and something being barbecued on it! I ended up putting pictures of trees on the walls to cheer up. You do miss the Earth there.'

There were a vast number of changes. Scott's DNA function altered, including increased activity of strands associated with fixing damaged DNA, possibly due to radiation. Kelly was exposed to 48 times more radiation than his brother back on Earth. His immune system changed; so did his concentration. One researcher said it was almost as if he got confused. There were changes to his microbiome as well, with new species of bacteria growing in his gut. There was some indication that some of the changes levelled off after six months in space, as if the body had adapted, though more research needs to be done. Interestingly, Kelly's telomeres got longer in space. These are chains of molecules at the end of strands of DNA. They get shorter with age and they are thought by some to be an important ageing process. It's not clear if the telomeres in his cells actually got longer or if a hidden set of stem cells started producing cells that had longer telomeres. In any case, it does not seem to be a fountain-of-youth effect as it wore off within 48 hours of his returning to Earth. If the effect does make you younger in some way, then it only works in space. Although average telomere length, global gene expression, and microbiome changes returned to near pre-flight levels within six months of Kelly's return to Earth, increased numbers of short telomeres were observed, and expression of some genes was still disrupted. But for some scientists it is the cognitive effects that provide most cause for worry.

Scott Kelly noted damage to his vision while in space. At first, he assumed the changes to be temporary but as other astronauts undertook longer and longer missions a pattern started to emerge. For Kelly, and most other astronauts, the changes gradually disappeared once the mission was over, but for others the changes seemed to be permanent. He says that when he flew his first mission in space in 1999, he didn't need corrective lenses, but during the mission he noticed things were getting blurry when he looked across the flight deck of the space shuttle. Back on Earth, his eyesight returned to normal. A few years later he started using reading glasses. On his second mission he noticed he didn't need glasses, but when he came back to Earth after a few weeks he needed them again.

Three years later, for his first long-duration flight, he was wearing glasses all the time and in space his vision got worse, and he wore stronger lenses. When he returned to Earth his vision went back to what it had been before he left. But his eye problems were not over. Kelly had swelling of the optic nerve and what seemed to be permanent folds in the region between the retina and the sclera – the white part – which provides oxygen and nutrients to the outer layers of the retina.

It's thought that as bodily fluids are redistributed in weightlessness there is an increased pressure in the cerebral fluid surrounding the brain that causes changes in vision. Astronauts say that the increased pressure in the head is something they learn to live with in space. 'We adjust over the first few weeks in space and pee away a lot of the excess, but the full-head sensation never completely goes away. It feels a little like standing on your head twenty-four hours a day – mild pressure in your ears, congestion, round face, flushed skin,' says Kelly. The curious aspect to this is that only male astronauts suffer such damage while in space. If the problem can't be solved, perhaps we might have to send an all-female crew to Mars.

In a write-up of the Kelly twins experiment, scientists noted that apart from the Apollo astronauts no one had ever been outside of the protective magnetosphere. It seems, then, that our bodies adapt to space after a few months, but not completely, and that the problems astronauts experience might increase with the duration of the mission. It's not looking good for Mars voyagers, though there is one obvious solution.

In the latter part of this decade, or the early part of the next, a long-duration mission will have to be performed at the Gateway, beyond the protection of the magnetosphere.

A crew of two will have to spend a year orbiting the Moon, taking blood samples every day along with many tests that will chronicle any degradation. The results of this mission could be what dictates that the first mission to Mars is an orbital rather than a landing mission. It would be a necessary test of the crew – and the spaceship.

The crewed spacecraft that will travel to Mars will have to be outfitted like no other spacecraft before it. Every other mission has an abort procedure – a way to get home swiftly. The Mars craft will not. As soon as its thrusters are fired and the velocity obtained to leave the Earth–Moon system (if the orbits are right it will only need a slight nudge) then very soon the abort options disappear and the craft and crew have no choice but to go through with the entire mission. All the food, fuel, oxygen and everything else they will need to survive they will take with them. For the voyage to Mars and back the only thing they will exchange with their home planet is information. So, how do they stay alive?

Every time a person breathes they exhale carbon dioxide. It's present all around us at trace levels but in the confines of a spacecraft it can accumulate if not removed. On the International Space Station this is done with a device known as the International Space Station Carbon Dioxide Removal Assembly (ISS CDRA). It's been called a finicky beast that requires a lot of care and attention to keep it working. If carbon dioxide levels creep up, it first causes mild discomfort for the crew. If levels increase a little bit more, they get headaches and start to feel congested. A little more and they start to experience mental impairment – in other words, they feel stupid. Repairing a CDRA is tricky. It has to be powered down and allowed to cool. Then all the connections — electrical, water and vacuum — have to be disconnected. Then it has to be unbolted and taken to pieces. In weightlessness it's not a straightforward task.

Clearly a Mars Transit Vehicle will need its version of the space station's CDRA, probably two. But is two enough? How often do they break down? What are the chances that two would be not working at the same time? Do you need three, or more? And what about spare parts? These are important questions because if the carbon dioxide cannot be removed from cabin air everyone dies.

Then the spacecraft will need a working toilet and waste disposal system. On the space station the water system is nearly a closed loop with only occasional need for a top-up, but top-ups aren't possible in interplanetary space. On the space station the urine processor distils the urine, turning it into drinking water. It's often broken on the space station and urine is stored in a holding tank until it's fixed. If the urine processor breaks down on the way to Mars, again, the crew die.

Maintaining complex hardware in space is much harder than it is on Earth. On the space station it's estimated that the crew spend a quarter of their time repairing things. If they can't they either use the 3D printer to make a critical component or ask Earth for a replacement part or entire unit to be sent up. If the Mars Transit Vehicle's 3D printer packs up they are in trouble.

The crew will have to be very carefully chosen, for they will be cooped up together in a confined space for years. As we have noted, long periods in space affect the brain more severely that we had thought. There is cognitive impairment, sleep disturbances, disorientation and altered perception. Problems are harder to grasp and solve, fellow crew members become more irritating. Then there is the stress, physical deprivation, limited space, noise, limited basic personal cleaning, constant fear of spacecraft or system failure and death, loneliness and isolation, perhaps leading to paranoia. The list of stress factors endured by the crew is long.

Astronauts are not robots and they have come into conflict with one another and with ground control. The crew of Apollo 7 in 1968 became very annoyed with their ground control and carried out a 'mutiny' in space. Despite their Apollo capsule being much larger than the previously used two-man Gemini capsules, their eleven days in space took their toll. It seems that once in orbit the more spacious cabin induced motion sickness, which had not been a factor in a smaller spacecraft. The crew were unhappy with the food, the cumbersome waste collection system, and then the commander Wally Schirra got a head cold and became irritable, frequently 'talking back' to ground control. In December 1973 the astronauts onboard the Skylab 4 mission conducted a day-long strike as a protest at being overworked. NASA had wanted to get the most out of the mission and gave then a packed timeline even telling them not to look out of the window at the Earth during their brief downtimes.

Space missions have been hampered by interpersonal problems. Many US astronauts experienced difficulties when staying on the Russian Mir space station as a 'guest' with no defined role. Other missions such as Soyuz 21 (1976), Soyuz T-14 (1985), and Soyuz TM-2 (1987) were shortened because of the bad mood and poor performance of the crews. Psychological factors also contributed to the early evacuation of a crew staying on the Salyut 7 space station.

In 2001, the National Academy of Sciences issued a landmark study titled *Safe Passage: Astronaut Care for Exploration Missions*. A panel of scientists from many disciplines identified some of the medical and behavioural issues that should be resolved quickly in anticipation of a return to the Moon and a mission to Mars. This far-ranging work covered astronaut health in transit to Earth orbit and beyond, health maintenance, emergency and continuing care, the development of a new infrastructure for space medicine, and medical ethics.

Some insights can be gained by looking at situations that have something in common with a flight to Mars. In 1992, France initiated plans for a new research station on the Antarctic Plateau, at a site called Dome C. They were later joined by Italy and Concordia Station became operational in 2005; the first 'winter-over' began in February 2005 with a staff of thirteen. Concordia consists of three buildings interlinked by enclosed walkways. Two large, cylindrical three-storey buildings provide the station's main living and working quarters, while the third building houses the electrical power plant and boiler room. The station can accommodate sixteen people during the winter and 32 people during the summer.

Dome C is one of the coldest places on Earth, with temperatures hardly rising above −25°C in summer and falling below −80°C in winter. The first summer campaign lasted 96 days, from 5 November 2005 until 8 February 2006, with 95 persons participating. The 2006 season included seven crewmembers with two medical experiments and the first of two psychological experiments sponsored by the European Space Agency. Two experiments investigated psychological adaptation to the environment and the process of developing group identity, issues that will also be important factors for humans travelling to Mars.

The largest study of the psychological effects of a Mars voyage was the Mars500 project led by the Institute of Biomedical Problems of the Russian Academy of Sciences in 2010–11. Six healthy male participants lived inside an enclosed module designed to give the feel of a Mars space-craft for 520 days. They had daily tasks to do and had a built-in delay of 20 minutes for communication to the out-side, simulating the time delay on a real mission. The mission was not an entire success. One crew member experienced depression and two others developed abnormal sleep–wake cycles and physical exhaustion. Another developed insom-nia. As the months dragged on, some crew members became increasingly sedentary while awake, requiring frequent rests. Over the course of the mission, the two crew members who exhibited the highest levels of stress were involved in 85 per cent of the perceived conflicts with other crew members. It seems that a single stressed-out astronaut can affect the entire mission.

But for some researchers, solving the problems caused by a lack of gravity outweigh almost all others and they advocate that the spacecraft has to have a component that produces artificial gravity. In other words, it has to spin. Many studies have been done on what size of wheel or arm would be best for a centrifuge on a spacecraft. Obviously, it cannot be too large, because it would require too much mass, nor too small because its dizzying effect would not be tolerated by the crew. So, what is the optimum size? Some believe the length of the spin arm should be about 10 metres, rotating at about five to six revolutions per minute. This means it would simulate lunar and Mars gravity depending on how far down the arm you go. But a centrifuge doesn't just simulate the force of gravity: it reveals other, rather disconcerting ones called Coriolis forces that can cause considerable disorientation.

Using a rotating structure would introduce major engineering challenges but it would counteract many of the long-term problems associated with weightlessness. It need not be used by the crew all the time. Experiments could be done to see if its effects are still worthwhile if astronauts only sleep in the centrifuge, in which case there need not be any windows causing the astronaut discomfort. Considering the intensive medical intervention and long periods of exercise the crew would have to otherwise endure, it seems like a simple solution, in principle. During a mission of stress and danger, closing your eyes to go to sleep and imagining you are back on Earth with its familiar gravity would be far better than taking drugs and floating around in a sleeping bag tethered to the wall, waking up to find your arms wandering in front of you.

If you examine what the astronauts on the International Space Station tweet about, it's often about homely things: their families and home towns, college football stadiums. It reinforces their ties to Earth. Going into space – be it to the space station or, especially, to Mars – is no business trip or visit to the other side of the world where you miss your home and family. The loss in going to Mars is the Earth itself: the beaches and the forests, the greenery and the deserts, the summer and winter, even the air. There is no substitute for those things in space, and their loss, for years, will be difficult to bear.

ENTRY CORRIDOR

———————•———————

This landscape was dead. It had never lived. It had been born dead when the planets first formed, a planetary stillbirth of boulders, coarse sand, jagged rock. The air was thin and so cold that it was closer to the vacuum of space than to any habitable atmosphere. Though it was nearly noon and the pallid tiny disk of the sun was high overhead, the sky was dark, the wan light shining on the uneven plain that was unmarked by any footprint. Silent, lonely, empty.

Only the shadows moved. The sun paced its way slowly across the horizon until it set. Night came and with it an even greater cold. Silently the dark hours passed, the stars arched by overhead, until on the opposite horizon the sun appeared once again.

Then something changed. High above there was a tiny flicker of movement as the sun glanced from some shining surface, a motion where none had existed ever before. It grew to a spot of light that blossomed suddenly into a long tongue of flame. The flame continued, even brighter as it came close to the surface, dropped, hovering. Dust billowed out, rocks melted and the flame was gone.

The squat cylinder dropped the last few feet and landed on wide-stretched legs. Shock absorbers took up the impact, giving way, then slowly levelling out the body of the device. It bobbed slightly for a few seconds and was still.

—HARRY HARRISON, *ONE STEP FROM EARTH*

The second ship to reach Mars, the John F. Kennedy which arrives in 2046, has been built with the experience of the James Caird II. It is larger, allowing its crew of six more space and more privacy. Two of the crew are medical doctors, specialising in space medicine as well as their other capabilities. The spacecraft has improved radiation shielding and a rotating sleeping section. It also has more advanced hybrid rocket engines.

You can get to Mars using conventional chemical rockets but there are potentially better forms of rocket propulsion. In a chemical rocket, a fuel and an oxidiser are combined, burning on contact, with the exhaust gases channelled into a nozzle to produce thrust. The vast majority of spacecraft have used such rockets.

Sooner or later nuclear-powered rockets will find their place in space. They could be twice as efficient. In the future they will be a propulsion option for travelling to Mars much more quickly than can be achieved by chemical means. One form is nuclear thermal propulsion (NTP) in which hydrogen is heated to extreme temperatures by a radioactive uranium pellet. The hydrogen is then ejected at high velocities, producing thrust. Such a rocket motor would be very versatile and could theoretically reduce the transit time to Mars to a few weeks.

One cannot overemphasise the importance of reducing the transit time to Mars, which using chemical rockets is 200–400 days. Such a duration requires a completely closed life-support system. Supplies of fuel, food and oxygen have to be taken in quantity. The length of time spent in interplanetary space increases the chances of encountering a solar storm, increases the risk of medical and psychological problems and requires ultra-reliable machines and adequate back-ups. But what if you could get to Mars in half the time or less? This would be possible with nuclear rockets.

Even before the first Apollo missions, NASA had its Nuclear Engine for Rocket Vehicle Application (NERVA) program in 1963 to develop such a rocket for interplanetary missions, but it was cancelled in 1973 because going to Mars was not politically popular at the time. In the past decade, interest in NTP has picked up and a study was carried out at NASA's Marshall Space Flight Center. It concluded that the Mars voyage could be reduced to three to four months. In May 2019, US Congress approved $125 million dollars in funding for the development of nuclear thermal propulsion rockets. Although this research doesn't have any part in the Artemis missions, NASA are being called upon 'to develop a multi-year plan that enables a nuclear thermal propulsion demonstration including the timeline associated with the space demonstration and a description of future missions and propulsion and power systems enabled by this capability'. The US Space Command in its requirements for the future wants a nuclear-powered spacecraft for use in cislunar space.

The efficiency of a rocket motor is measured by what is called its specific impulse, which rates how effectively it uses propellant, specifically the change in momentum achieved per unit of propellant used. The greater the specific impulse, the more efficient the rocket and the better its performance. It's usually expressed in seconds. Let me give you some examples. The solid-fuelled rocket boosters that were used on the side of the space shuttle during take-off and the early phase of ascent has a specific impulse of 250 seconds. A standard liquid oxygen–liquid hydrogen rocket has a specific impulse of 450 seconds – much more efficient. An ion thruster such as that used on Smart-1 comes in at 3,000 seconds, and a nuclear thermal rocket could be from 3,000 to 10,000 seconds.

After years of neglect, a revival of interest in this technology does seem to be happening, partly as a result of military interest. It is therefore possible that when human missions are flown to Mars, in twenty years or so according to my predictions, they would use chemical and electric/nuclear propulsion that would shorten the flight time to six to eight weeks. I hope so because it would reduce the risk considerably.

As we have seen, landing on Mars is a very different landing to that on Earth. Our thick atmosphere gives the crew more control and longer times to act. A spacecraft can be slowed to Mach 1 while still about 20 kilometres above the ground just by using a heat shield. This can't be done in Mars's thin atmosphere.

When the crew arrive at Mars, they face the most difficult part of the mission. This must be done in the face of a variable atmosphere with winds and turbulence and they must land at a particular place with high precision. I have seen many presentations about going to Mars and I have come to expect that the landing problem is treated as straightforward, which it is not. For unmanned spacecraft, mission controllers often say that a Mars landing is Mission Six Minutes of Terror.

Solutions to the problem depend upon what type of spacecraft you want to use to land on Mars. The Boeing company has a design for a large spacecraft that will have thermal protection and enough thrust capability to have a successful powered descent. It's going to take a lot of design, testing and expense to get such a vehicle to work, but if it can be made to work it is a good solution.

Another option is to send the landing craft to Mars before the crew so that when they arrive in the Mars Transit Vehicle they will have to rendezvous and dock with it. There are pros and cons to this approach. It will involve an extra launch to Mars, presumably when the Mars Ascent Vehicle (MAV) is sent. The basic plan is that the means to get back into orbit is already ready and waiting on Mars before the crew even set off. The MAV would stay on the surface for perhaps several years, automatically extracting oxygen from the atmosphere while releasing drones and rovers to look at the local area for signs of water ice near the surface. The preplaced MAV is regarded as essential, but the orbiting lander is still being evaluated. It will probably have to wait in orbit for years and then when the crew arrive the rendezvous and docking will be a manoeuvre that will have to be performed successfully, otherwise no landing.

When the computer confirms that the lander is in the correct 'entry corridor' the crew would perform the de-orbit burn at an altitude of 400 kilometres – the most dramatic 400-kilometre journey of their lives. They will be 30 minutes from touchdown if all goes well. Twenty minutes later they will have descended to 130 kilometres, where their craft will be at what's often called the interface, the point where they start to feel the atmosphere. They are then ten minutes from touchdown, and things start to happen quickly. When they reach 70 kilometres the deceleration is equivalent to the Earth's pull of gravity and they will briefly experience this sensation for the first time in a year. The deceleration has to continue beyond this, and in Boeing's design this means changing course to lose speed, flying in the shape of an 'S' to dissipate velocity before turning tail down for the final powered descent.

Whatever design of lander is used, its forward speed will be much greater than its rate of descent. We can imagine the crew of the John F. Kennedy in their lander, the ground below moving beneath them at 100 kilometres a minute. The rocket motors are pointing ahead. It is a very dangerous time. With five minutes to go, the lander is experiencing its maximum heating as it travels at six kilometres a second. The G-forces reach their maximum. The retro-propulsion rockets fire. The computer enters its fine guidance mode and seeks the best landing spot.

In the final minute the crew see the MAV and the other equipment pre-positioned around it. The lander must not set down too close as its exhaust will kick up rocks that could cause damage.

When it comes to choosing a landing site for humans there are two views: go for the ice, which means mid-latitudes, or go for a thicker atmosphere, which means low points nearer the equator. Equatorial sites have relatively moderate and stable temperature changes with none of the extremes seen at higher latitudes. They have consistent 'high-angle' light all year round, which will be good for solar panels and habitat illumination. Best of all, we have good on-the-ground data for such sites. Seven of the nine success-ful landed missions have visited equatorial locations and all three rover-only missions have operated in equatorial zones, for durations of between six and fifteen years, recording tem-perature, atmospheric dust content, pressure, humidity and radiation.

For advocates of low-latitude bases the plan is go as low as you can. Because the Valles Marineris is among the lowest places on Mars, about ten kilometres below Mars's average level, the atmospheric pressure is higher than normal. In fact, it could at certain times be above the triple point of water, which means that for some periods during the day water could exist on the surface without evaporating. That might prove useful. It might also be a good place to look for life. Additionally, the extra kilometres of atmosphere would be useful for a spacecraft when landing and it would also confer some additional radiation protection.

So what about landing in the Valles Marineris? Such a site would certainly provide a lot of scientific diversity. The suggested Noctis Landing zone is a regional depression at the western end of Valles Marineris dominated by hydrated minerals and with many indications of having spent a long time underwater in the past. Spectral analysis from orbit indicates the possible presence of actual water in the area on occasions throughout the year.

Others place a premium on the ice: go to where we know there is abundant ice – and that means higher latitudes. If such a landing site is chosen, the first crew will be preceded by an unmanned craft that will extract liquid oxygen from the air and drill into the ice to produce oxygen and hydrogen. The landing crew will not leave Earth until they know that the resources they will need have been made and are safely stored.

Making a base on Mars will be similar to making a base on the Moon: temporary shelters at first and 3D formed habitats later. In 2018 NASA selected the winners of a competition to design a Mars base. They included a community of modular pods and a tall, egg-like, habitat. Arkansas-based team Zopherus won first place. Its plan was to have a lander scan its surroundings and select an appropriate area, while autonomous robots are deployed from it to gather materials for the 3D printer. The ground has to be sealed in the same way that lunar platforms are. The 3D printing robots will mix the regolith and begin to print the hexagonal structure. Being hexagonal, the modules can nestle close to each other and be connected by tunnels. In the multi-floored design there is a mezzanine level on each floor with windows providing extensive views of the landscape as well as allowing light to enter to grow crops. When the competition went to its second phase, which involved detailed three-dimensional digital templates that could be fed directly into 3D printing software, two other entries, from SEArch+ and Apis Cor, took first prize with their design of a habitat for a crew of four to live in for a year.

Drilling into the ice on Mars is a technology that can learn a lot from those industries that have to drill into Earth's polar regions. There are two proven terrestrial technologies that could be adapted: coiled tubing (CT) for drilling and the Rodriguez Well (RodWell) for water extraction. CT rigs use a continuous length of tubing that is flexible enough to be wound on a reel and rigid enough to withstand drilling forces and torques. The tube is pushed down a hole using so-called injectors – actuated rollers that pinch the tube and advance it downwards – and at the tube's end is a motor and a drill bit. A hole is drilled by advancing coiled tubing deeper into the subsurface while blowing cuttings out of the way. A commercial CT rig weighs fifteen tonnes and drills to 500 metres at 1 metre per minute in hard rock. RodWell is a technology where a hole is drilled in ice, which is melted and pumped to the surface. It has been developed and tested in Antarctica in the 1960s and regularly used at the South Pole.

At some point, after the first few missions, the crew are going to have to grow food on the Moon and Mars. It's obviously more important for Mars, given its distance and the limited nature of the supplies from Earth. The margin between life and death on Mars is far narrower than on the Moon. From what we know, the Martian regolith contains many of the essential nutrients needed to support plant growth. However, it may also contain many substances that will harm anything that tries to grow in it. It has high salinity and contains toxic perchlorate and heavy metals. In addition, it seems to have low water-holding capacity and obviously an extremely low availability of organic matter. This would have been bad news for Mark Watney in *The Martian*. Before he could spread Martian soil inside his habitat to grow potatoes, he would have to spend considerable effort and resources in washing it.

Experiments have been carried out using simulated Mars soil (you can buy it — it's called Mojave Mars Simulant, developed at NASA's Jet Propulsion Laboratory) to see if it can be treated to make it more hospitable for plant growth. Microbial inoculants including plant-growth-promoting bacteria known to improve plant response in salty conditions have been added and the regolith structure altered to see if it improves things. But the worth of such experiments is seen in the crops that can be grown in treated Mars soil. Lettuce, spinach, tomatoes, peas, leeks, radishes, red clover, moth bean and salt grass have all been tried. In some experiments spinach did not grow well, neither did peas or potatoes (sorry Mark Watney). The winners were basil, kale, hops, onions, garlic, lettuce, sweet potatoes and mint. The researchers think coffee has potential. Hops seemed to do very well, so if we ever get to Mars there could be beer. It's a start.

By the 2030s China will send spacecraft to the planets, possibly making more use of AI systems than other nations, and it will have its Moon base towards the latter 2030s. But what will China do with Mars?

In all likelihood the country is going to get richer, and if it so decides, the resources for space exploration could be considerable. It has a slow and deliberate space program, making advances, learning from them and from other nations, and moving on to the next goal. In the next decade it will have a small but growing space station. Perhaps it will have the only space station in orbit – as by common view the International Space Station, from which China is banned, is nearing the end of its useful life. It was given an end date of 2024 but it might survive a few more years before it is decommissioned by de-orbiting it into the sparseness of the South Pacific.

Some believe that a US–China rivalry in space would be a good thing. Two missions to Mars could involve cooperation. I can't quite see that everyone would get together and have a joint mission to Mars. I don't think the national mood in the US or China would want that. But competition is not a bad thing. A space race to Mars? History has done a similar thing before, and didn't Neil Armstrong say that nations didn't go to the Moon just for science or even just for the sake of exploration. A necessary ingredient might be rivalry. Perhaps China will be the nation that develops many bases and perhaps it will be the first to explore the caverns on the flanks of Olympus Mons, using experience gained living in lunar lava tubes. Inside those caverns of Mars, it will be warmer because of the volcano's heat; it will be protected from radiation and have a more equitable climate. Perhaps that is where we could look for life.

PART 3

AND BEYOND

DOWN TO
A SUNLESS SEA

———————•———————

Beyond Mars there is Jupiter. Ancient peoples all over the world knew Jupiter. Only Venus is brighter but Venus, being closer to the Sun than the Earth, never leaves the evening and morning skies. This wandering lordstar was identified as the most powerful of the gods as it rules over dark midnight skies. Only six probes have ever been there; only two have orbited it. Humans will not reach that far in the next 50 years, and perhaps not in the next 150. It would be a gruellingly long flight and on arrival the intense bands of lethal radiation around Jupiter would have to be avoided.

It is the largest and most dramatic of all the planets, eleven times the diameter of the Earth and with a mass over twice as much as all the other planets put together. It is a gas giant world without any visible solid surface; instead we see a multi-coloured atmosphere with bands, jet streams, cyclones and anticyclones and even a storm the size of the Earth – the Great Red Spot – that has probably raged for millennia. Its four major moons are each dramatic in their different ways. They are called Io, Europa, Ganymede and Callisto, names suggested by Simon Marius, a contemporary and rival of Galileo.

Pioneers 10 and 11 reached Jupiter in November 1973 and November 1974 respectively. When 6.4 million kilometres from the planet, Pioneer 10's magnetometers detected the bow shock, the region where the solar wind encounters the Jovian magnetosphere and pushes it back. In December it made its closest approach, sending back pictures taken by a simple line-scan camera that were three times better than any taken from Earth. On 3 December it passed within 130,000 kilometres of Jupiter's cloud tops. Pioneer 11 emerged from the asteroid belt the following spring and, encouraged by the success of Pioneer 10, scientists decided to send it even closer to Jupiter – its aim point just 34,000 kilometres above the clouds. The radiation dosage taxed the spacecraft to the limit.

Voyager 1's close encounter in March 1979 took place over little more than 48 hours from the inbound encounter to the outbound crossing of the orbit of Callisto, just a month after its pictures became better than those obtainable from Earth. Arriving in August that year, Voyager 2 took a complementary trajectory. After its closest approach to Jupiter, a ten-hour watch of Io was performed.

Io, to everyone's surprise, turned out to be the most volcanically active world in the solar system, Earth included. But its volcanoes aren't eruptions of liquid rock – magma – as takes place on Earth. They are cold sulphur plumes, blown

250 kilometres above the surface, ejected at rates of 1 kilometre per second (ejection velocities from Mount Etna have been measured at 50 metres per second). Liquid sulphur may flow considerable distances over its surface, obscuring any features and coating this tiny globe in red and yellow. Eight active volcanoes were seen, although it's estimated that Io has on its surface about 200 volcanoes (geysers is a better term) larger than 20 kilometres. The Earth, with three-and-a-half times more area, has only about fifteen in that size range. These alien volcanoes are dramatic manifestations of tidal energy dissipated inside the moon.

Europa, the smallest of the Galilean moons – that is, those seen by Galileo – showed a number of intersecting linear features that were first thought to be cracks caused by movements of its crust, but closer pictures left scientists puzzled. The streaks seemed to be flush with the surface, possibly because the icy surface was fractured and then filled with material from below, making Europa the smoothest world in the solar system. Very few impact craters were seen.

Ganymede, the largest moon in the solar system, showed two distinct types of icy terrain – cratered and grooved. Callisto, the outermost of the Galilean moons, had a surface that was very ancient and the most heavily cratered of all of them – almost as heavily cratered as the highlands of our own Moon. There were remnant rings of large impact basins, parts of which had been erased by the slow, elastic creep of the icy crust.

During Voyager 1's passage through the equatorial plane, seventeen hours before closest approach, a ring of fine material was found, no more than 50 kilometres thick, narrower than Saturn's rings and made of smaller particles, orbiting 128,000 kilometres from the centre of the planet. A small moon, perhaps 20 kilometres across, was discovered by Voyager – in fact the first ever moon discovered by a spacecraft. It was named Adrastea and it skirts the periphery of the ring. The slightly larger moon Metis orbits inside. They were called shepherd moons because they influence the orbit of the ring particles.

Since Voyager, only two more probes have visited Jupiter. The Galileo probe orbited it for almost eight years from 1995 and made many discoveries. It confirmed the idea that beneath Europa's icy shell lies an ocean of warm water. One scientist said that the last time we discovered an ocean it was the Pacific. In 2016 the Juno mission arrived and continues there still. From its orbit it has changed our understanding of Jupiter's polar regions sending back some spectacular images.

Europa's subterranean ocean has become an obsession for some. It's there beneath Europa's frozen landscape because there are titanic energies at work. Tides raised by Jupiter's gravity cause the surface to rise and fall by 20 metres every three days. As the ice and rock flex, their movement pumps energy into its interior, which is hot and keeps the under-ice ocean warm. It is a place protected from the vacuum of space by a layer of ice more than 10 kilometres thick. It has been a stable environment for possibly tens of millions of years.

Plans to return are underway. The Europa Clipper spacecraft is due to be launched in 2024 and it will arrive around 2030. It will not orbit Europa as the radiation environment is too harsh for the spacecraft to survive very long. Also, the presence of Jupiter makes any orbit around Europa unstable. Instead it will go into orbit around Jupiter, making up to 44 brief flybys. Even so, the craft will last only a few months, even with its 150 kilograms of titanium shielding. There was at one time to be a more ambitious Jupiter Europa Orbiter mission but the price tag of $4.5 billion became too great to bear. Scientists believe that by performing flybys the Europa Clipper can do almost as much at half the price.

The JUICE (Jupiter Icy Moons Explorer) mission will also be heading for a Jovian moon, this time Ganymede, which also seems to have an under-ice ocean. Although it will reach Jupiter in 2029, the craft will have to undertake a complex series of flybys and gravity assists before it can enter orbit around Ganymede in 2032, becoming the first spacecraft to orbit a moon other than our own.

JUICE and the Europa Clipper could pave the way for a Europa lander that would operate for 20 days or so on the surface. This project is still seeking funds but if launched in 2025, or more likely nearer 2030, it could set down in the late 2030s. But even a landing among the ice groves and spires of Europa is not the ultimate mission. Sometime in the next 50 years we will have accumulated the technology to attempt a mission to not only land on Europa but drill through its icy crust to burst into the sunless sea that lies beneath it. For some mission engineers it's the only mission that in their heart of hearts they want to tackle in their lifetimes.

In 1974 a survey aircraft was flying over central Antarctica, its course set for the desolate and little-known central polar wastes. Radar was used to measure the profile and properties of the ice beneath. After a flight of a few hundred kilometres, the survey team detected a very peculiar signal – a double echo. What they had seen was the radar signal of liquid water trapped 4 kilometres down. Years later, they went back to take a closer look and realised they had found a subterranean ocean 19,000 square kilometres in size, twice the size of Yorkshire: a cold under-ice ocean cut off from the rest of the world for at least 2 million years. Sounds like Europa.

Lake Vostok – named after the nearby Russian research station – was formed when geothermal heat from below melted the ice above. Microbes and other forms of life may inhabit it. Sediments deposited at the lake bottom by the glacier as it grinds over the bedrock would be a possible source of nutrients. In this closed ecosystem something may have evolved that's different to what's been found elsewhere. Some call Lake Vostok the Europa simulator – it's the largest of almost 400 known subglacial lakes.

The ice bubbles and deposits laid down during the accumulation of ice means that Lake Vostok's overlying ice provides a continuous climatic record of the past 400,000 years, although it's estimated that the actual lake itself may have been isolated for 25 million years. In February 2012, a team of Russian scientists drilled the longest ever ice core of 3,768 metres and pierced the ice shield, reaching the lake's surface. The first core of freshly frozen lake ice was obtained in January 2013 from a depth of 3,406 metres. Unfortunately, as soon as the ice was pierced, water gushed up the borehole, mixing with the freon and kerosene used to keep the borehole from freezing. A new borehole was drilled and what is believed to be pristine Vostok water was obtained in January 2015. In the future it's planned to lower a probe into the lake to collect water samples and sediments from the bottom.

The goal of a future Europa mission is to get to the moon's under-ice ocean, which could be at the same depth as Lake Vostok. After landing, the core of the probe would heat up and begin to descend into the ice, boring a hole just 50 centimetres across. It will trail an umbilical behind it as once the probe disappears, the water refreezes almost instantaneously. Emerging hours later into the sea beneath the ice, the probe pauses and turns on its lights. The first task is to wait and listen. Sensitive microphones strain for the sounds of this alien ocean, the creaking of the ice above and the 'ping' of an acoustic sounder.

A series of electronic tongues taste the water and provide a rapid chemical analysis. Two searchlights, one narrow to look as far as possible, the other a wide beam, illuminate the dark, the first bright light in this ocean for tens of millions of years, perhaps much longer. A smaller sub-probe is deployed and darts away, straight down, trailing a fibre-optic cable. For long minutes it continues to fall. Then rocks come into view. Reaching the sea floor is only half of its journey. It now has to find a hydrothermal vent. Its sensors have already detected a chemical gradient and it moves in the direction of it, following a current of sulphur compounds. Perhaps these hydrothermal vents and the minerals that are deposited around them, accumulating over the ages, are enormous, towering above the ocean floor like the skyscrapers of New York, each with its own colonies of microorganisms fed on by alien shrimps and fishes. I wonder if as you read this the forces of evolution are in action on this tiny moon a thousand million kilometres away.

I have spoken to many engineers about this mission and all know it is way beyond us at the present time. How do we make a probe to melt through kilometres of ice that is so cold it is like bulletproof steel? How do we get the data from the bottom of the ocean to the surface of the ice? When can we do it is a question answered by a shrug of the shoulders; 2050s, someone says, possibly? But oh, what a mission.

But what of Io, which to many is the most amazing moon of all. There are no missions funded but sometime in our 50-year time frame the Io Volcano Observer may approach its target with its hyperactive geology. The tidal forces between Io and Jupiter generate a large amount of heat, keeping the moon's subsurface crust liquid. The surface is constantly renewing itself, filling in any impact craters with molten lava lakes and spreading liquid rock over the landscape. We would like to know the composition of the surface as well as more about its magnetosphere. As Jupiter rotates, it sweeps its magnetic field past Io and strips off about a tonne of material a second which forms a doughnut-shaped cloud of intense radiation called a plasma torus. Some of the particles are pulled along lines of magnetism into Jupiter's atmosphere, creating auroras in the upper atmosphere. Io's passage through the Jovian magnetic field turns it into a generator creating an electric current of 3 million amperes. This current creates lightning in Jupiter's upper atmosphere. Scientists would love to take a closer look at this process, though designing a spacecraft to venture into these regions of titanic energy would not be easy.

If its moons are dramatic then Jupiter's atmosphere is scarcely less so. The Great Red Spot, big enough to swallow three Earths is a high-pressure area circulating counter-clockwise, taking about six days to complete one rotation. When the Galileo orbiter was scrutinising Jupiter it released an entry probe that plunged into Jupiter's clouds in December 1995. As it descended through the cloud layers it sent back data for 57 minutes.

What about going back, this time to drift among Jupiter's wispy, white ammonia cirrus. To descend through sulphur stratus through clear lanes in the clouds down to deeper clouds of water. It could also rise to observe the cloud-top lightning bolts and even set sail to the poles to study the aurorae. The atmosphere is some 1,000 kilometres thick, though the region where there is weather is only 80 kilo-metres in depth. There are cyclones and anticyclones, jet streams, zones, spots and belts, looking more like swirling colours on an artist's canvas than storm systems the size of worlds. A Jovian Buoyant Station would be unlike anything we have ever done before. We must go there.

TIGER STRIPES

Everyone knows it's the spectacular rings that give Saturn its majesty. Galileo observed the planet with his crude telescope in 1610 but didn't make out the rings, suggesting that Saturn wasn't just one planet but three! Later he said it looked as if the planet had 'handles'. It was Dutchman Christiaan Huygens who in 1655 realised that a ring system encircled the planet. Wonderful world that it is on its own, Saturn has two gems nestling in its orbital arms, the moons Titan and Enceladus.

Titan is the second largest moon in the solar system after Ganymede and is the only moon to have a dense atmosphere. We have sent a probe through Titan's thick clouds down to its oily surface. It was the Huygens probe of 2005, which hitched a ride on the Cassini Saturn orbiter. As it parachuted to the unknown surface, many wondered: would it land in mountains, on a plain or in an ocean? As it looked down it seemed to be headed for a shoreline. What seemed to be drainage channels were visible. At the landing site there were indications of pebbles of water ice coated by a thin film of methane. It seemed to be a dried-up region that had recently seen a liquid of some sort. As it passed by, Cassini detected lakes of hydrocarbons, making Titan the only body in the solar system other than Earth that has stretches of liquid on its surface. Huygens lasted for 30 minutes.

Titan's atmosphere is 90 per cent nitrogen, 6 per cent methane, and 4 per cent argon. The atmospheric pressure is 1.5 times that of Earth. The surface gravity is one-seventh that of the Earth, slightly less than the Moon. Beneath its clouds there are deserts with dunes of water ice crystals and mountains made of hydrocarbon ice; there are networks of liquid methane: rivers, lakes, and seas. In the mid-latitudes there could be liquid methane swamps. Standing on Titan you would see a dense orange haze while chemicals called tholins fell from the sky – the product of sunlight falling on methane or ethane. They may have also once rained down on the primitive Earth and played a part in the origin of life.

There could be a methane cycle on Titan, meaning that it might rain droplets of methane. If it does, would there be rainbows? If so they would be larger than those seen on Earth because of the different refractive index of liquid methane compared to water. They would have a primary bow radius of 49° as opposed to 42° for Earthly rainbows. The order of colours, however, would be the same: blue on the inside and red on the outside, with an overall hint of orange caused by Titan's orange sky. The problem is that rainbows need direct sunlight and Titan's atmosphere is distinctly hazy. Measurements from the Cassini and Huygens probes show that the atmosphere is transparent to infra-red light, leading some scientists to speculate that infra-red rainbows could be quite common on Titan.

The next planned mission to Titan is called Dragonfly and will reach it in 2034. Once it reaches the surface it will take off again, using a rotor to ascend, and make its way across a shoreline. It's very easy to fly on Titan: the gravity is weak and the atmosphere is thick, so lift is easily achieved. For the future scientists will want to look at the oceans of liquid ethane or liquid methane. A boat or even better a submersible sailing on the methane waves of the so-called Kraken Mare would be interesting.

But Saturn has an even tastier place to explore.

Sometime in the next 50 years, swooping low over a lonely ice-moon of distant Saturn, an unmanned spacecraft will manoeuvre carefully. Under the command of an onboard computer that received its last orders from Earth days ago, it races over a shining landscape of ice and rock, picking out landmarks and waypoints as it closes on its target. Looming above it is mighty Saturn, a huge presence, 30° across, dominating the sky. Its magnificent ring system, edge-on from this viewpoint, appears as a faint, thin band of light. Its true extent is revealed by the huge shadows it casts over Saturn's southern hemisphere. Another of Saturn's moons, Mimas, is beginning a passage across the face of the planet, and not far from it is the fuzzy, orange blob that is Titan, ten times larger than the tiny world the probe is exploring, which astronomers call Enceladus. Small it may be, only about the size of England, but for some it offers the most promising place to look for life beyond the Earth, and some conceive, as yet just in their imaginations, of missions that would reach this icy world and search for its secrets.

The probe's cameras record high-resolution images as it fires laser pulses to bounce off the surface, its travel time measuring the topography of the changing textures passing below. Flying sometimes over flattish, uncratered regions like the Sarandib and Diyar Planitia, sometimes over grooved terrain like Harran Sulci, ridges like Cufa Dorsa and trenches like the Daryabar Fossa, the few craters the probe encounters are fractured and deformed. Soon the land changes and becomes noticeably smoother, signifying that it has reached one of the most significant places in the solar system. The probe is nearing the so-called 'tiger stripes' where there is mystery and possibly life, and, on a world where there is almost no atmosphere, a strange mist looms on the horizon.

Little was known about Enceladus until the Voyager probes arrived in 1980 and 1981. It was a big surprise. Much smaller than our moon, it had a surface of rock and ice with few impact craters. It was not battered and scarred like many other worlds of the same size: instead it had a youthful appearance. Clearly, something was going on at Enceladus. The more sophisticated Cassini space probe arrived at Saturn in 2004 and took orbit of the planet, making several close flybys of Enceladus, and with each one it became more interesting.

Its sheen of ice made it the most reflective world in the solar system. Icy moons are not unusual out there in the solar system's cold, outer reaches, but a closer inspection of Enceladus revealed something unique. Across its northern ranges are layers of ice through which are exposed rocky mountain folds. In the southern regions the surface is younger and deformed by crustal movements as if the ice and rock moves like a viscous liquid taking millions of years to flow. The surface is young, in places just 100 million years old, in other places perhaps less than a million. For the surface of a world in our solar system this is young indeed. The 'tiger stripes' are four almost parallel low ridges, each with a central fracture straddling the moon's south pole, winding over a heavily deformed region. They are 32 kilometres apart, 130 kilometres long, 2 metres wide, 0.4 kilometres deep, with 100-metre-high flanking regions. Cassini's infra-red spectrometer picked them up as being slightly warmer than the surrounding terrain, giving off an estimated 3–7 Gigawatts of power.

But it was when Cassini looked at the limb of the moon that scientists got the greatest shock. Silhouetted against the darkness of space were jets of what turned out to be water vapour and ice, and they were coming from the tiger stripes, which were obviously cracks in the surface of Enceladus, reaching down to an ocean beneath the ice. It seems that the ice beneath the moon's icy skin has melted because Enceladus is made to flex due to strong tides raised inside it as it is held in a so-called gravitational orbital resonance by the nearby moon Dione.

In July 2005 Cassini flew through the plume with its spectrometer and dust analyser working overtime. The plume was shooting up from Enceladus at more than 3,000 kilometres per hour from pressurised underground water caverns. The ice and dust escaped into space and went into orbit around Saturn forming the so-called e-ring of fine debris that circles the planet. Cassini also found a unique chemistry.

There was salt water and carbon dioxide as well as more complex hydrocarbon molecules such as propane, ethane and acetylene. Such molecules could be the building blocks of life. There are also indications from the plume that it originated from very hot water.

As we have seen with Europa, scientists have become intrigued by the possibility of life in oceans trapped beneath the icy crusts of distant worlds. But the creatures of those Jovian worlds, if they exist, lie trapped beneath many kilometres of super-cold ice. Only at Enceladus does there seem to be a way to get to such an under-ice ocean. Nowhere else do we know of a such a cracked world. If we can really get in through those cracks, we could be witnesses to the first aliens we have ever encountered.

Despite its scientific importance and the mystery of its leaking ocean there are no definite plans to return to Enceladus. A probe to return to Earth a sample of the plume material would be scientifically invaluable, as well as technically challenging. Even more ambitious would be a landing. Sometime in the next 50 years a tractor could make its way to the rim of one of the Tiger Stripes and lower a probe into dark, misty ice caverns, and watch the pools of water flow over the floor, paying particular attention to the curious slime at its margins.

Sometimes I wonder if Mars is the best place in the solar system to look for life. For years the search for life on the red planet has been accompanied by the mantra *follow the water*. Mars has little water but a lot of ice. The water is found beneath the icy crusts of smaller worlds farther out in space. Perhaps that is where the life is.

PILOTS OF THE
PURPLE TWILIGHT

———•———

Exploring distant parts of the solar system will surely lead to more interest in developing new, more efficient means of getting there. Sometime in the next 50 years we could start using the power of light to travel across the solar system. I imagine a spacecraft with a folded wing of ultra-thin reflective metal nestling behind a solar shield. When launched, it would approach the Sun, possibly assisted by a flyby of Venus. When close to the Sun, where the radiation is intense, like a butterfly beginning a terrestrial flight it will discard its sunshield and unfurl the micro-thin sheet, a few kilometres across, and emerge into the solar glare. Now brighter than any star, the sheet strains and billows. The pressure of sunlight is filling it as it opens and takes flight. Like the seeds that leave a dandelion propelled by human breath, it heads for the outer planets.

Light applies a slight pressure on any illuminated object. In 1924 space pioneers Fridrikh Tsander and Konstantin

Tsiolkovsky noted that in the vacuum of space a large, thin sheet of reflective material could be a propulsion device requiring no propellant. Sunlight would push it along. For a highly reflective sail the solar flux could produce a force of about 9 Newtons for every square kilometre (a Newton is the force needed to give a mass of 1 kg an acceleration of 1 metre per second every second), which is quite useful since it is continuous.

In 1973 NASA sponsored a study for a solar sailing probe to intercept Halley's comet, pulling behind it a platform of scientific instruments. A team from the Jet Propulsion Laboratory (JPL) designed a number of solar sails. The recommended JPL sail craft had a central mast and booms that would spread and support a 2-micron-thick plastic sheet coated with aluminium 850 metres square. Making one side of the sheet reflective and the other dark would set up a force imbalance that could be used for additional manoeuvring. It all weighed 5 tonnes, needed little highly advanced technology and had to be launched in 1985 to meet the comet's passage the following year. However, it was decided that the Halley's comet interceptor should have a different form of propulsion called a solar-powered electric propulsion system. Unfortunately, both studies were soon rendered academic when the project was cancelled by the US Congress.

The idea did not die. In 1979 the World Space Foundation, a non-profit organisation of space enthusiasts that included many JPL scientists, took it up. In 1981 the world's first solar sail, a half-scale prototype, was exhibited. Two years later a full-size prototype was completed.

In the US the Planetary Society has championed solar sailing for several decades. In 2005 they attempted to launch the world's first solar sailing spacecraft, Cosmos 1, but its Russian Volna rocket malfunctioned and it failed to reach orbit. NASA tried to launch a technology demonstration mission for solar sails in 2008 on a Falcon 1 rocket, which unfortunately also failed to put the spacecraft into orbit. A few years later they tried again with their NanoSail-D2 mission made from spare parts from the first spacecraft. This time it did get into space. When fully deployed, NanoSail-D had a surface area of almost 10 square metres and was made of CP1, a polymer no thicker than single-ply tissue paper. The first big challenge for researchers was to pack it into a container smaller than a loaf of bread and create a mechanism capable of unfolding the sail without tearing it. This was done successfully but the results of the mission were rather inconclusive. Then there was the Sunjammer, an ambitious mission slated for launch in 2015. When the sail was being made in 2013 it was the largest ever constructed, but the project was cancelled before it was launched. It was named after a 1964 story by Arthur C. Clarke in which solar sail craft compete in a race across the solar system.

In 2015 the Planetary Society tried again with their LightSail 1 spacecraft that successfully completed a test flight. In June 2019 an improved version, LightSail 2, was launched and deployed its solar sail the following month. Shortly afterwards mission controllers adjusted LightSail 2's orbit, demonstrating that solar sailing is a viable means of propulsion for small satellites at least. There has also been the Japanese Aerospace Exploration Agency's IKAROS spacecraft, which first demonstrated controlled solar sailing in interplanetary space, using a sail 20 metres across. Seven months after launch it flew past Venus.

The Near-Earth Asteroid Scout (NEA Scout) is a planned mission by NASA to use a small solar sail to carry a very small satellite called a CubeSat. In recent times such satellites have flourished: usually doing a single task, they are straightforward to build as they use standardised components, and being small and light can hitch a ride into space on a rocket for a modest price. NEA Scout wants to send its solar-sail-powered CubeSat to a near-Earth asteroid. It will be one of thirteen CubeSats to be carried with the Artemis 1 mission into a heliocentric orbit in cislunar space on the maiden flight of the Space Launch System in 2021. It will carry out a series of lunar flybys before beginning its two-year-long cruise to an asteroid. The particular target has yet to be decided. In the future, CubeSats will be sent all around the solar system and prove themselves very useful, gathering data on almost every type of object.

The goal of the NASA mission is to show that CubeSats powered by a solar sail can visit an asteroid at low cost, especially near-Earth asteroids in the 1–100 metre size range, as we could do with knowing more about them. Should we ever be threatened with the impact of such an object, we had best know how many of them are solid chunks of rock and how many could be relatively harmless rubble piles.

Solar sailing spacecraft could perform in orbits that are difficult to reach by conventional spacecraft because they can use the sail's acceleration as a balancing force. They could be used for solar monitoring missions positioned between the Earth and Sun at a closer distance than would otherwise be possible.

A 4-kilometre-square sail deployed in the vicinity of Earth could transport a payload to Mars and then return to Earth for more. A conventional spacecraft powered by chemical rockets would get there sooner but would need two to three times more fuel than cargo. Perhaps six of these Sunjammers could form a bridgehead to Mars.

Sunjammers could travel even further. Voyager 2 will take some 80,000 years to travel a distance equivalent to the 4.3 light years (about 40 million million kilometres) to the nearest star. A conventional solar sail could perhaps reduce this to about 15,000 years, still almost an eternity when compared with our individual lifespans. However, some scientists have suggested that we may be able to help Sunjammers along by firing lasers or microwave beams at them. Such ideas, although fascinating, are speculative, but the concept is simple. Is it too fanciful to think of future space mariners plying the trade routes between the planets on sails of gossamer that ride the sunlight? Some calculations suggest the flight time to Alpha Centauri would be about 500 years.

The Breakthrough Starshot mission program, initiated in 2015 by Russian billionaire Yuri Milner, proposes to make use of advanced microelectronics to create a probe with a mass of just 100 grams, sail included. A 24-megawatt laser would blast them on their way. Some argue that such a combination of microspacecraft, laser, and sail technology could well become available within the next decade or two, putting a precursor mission for interstellar exploration on the agenda for the next 50 years.

The challenges for an interstellar flight are enormous. The distances dwarf anything else we have done. We have been to Pluto, which is on average almost 40 times further away from the Sun than is the Earth, or 40 AU. The nearest star system — Alpha Centauri and its companions — is 270,000 AU away. The fastest spacecraft we have ever launched is the Parker Solar Probe at 246,960 kilometres an hour. At that speed it would take many hundreds of centuries to traverse that distance. The basic numbers associated with interstellar distances are daunting.

But still it's a dream for some.

In 2014 there was a competition to design a spacecraft able to reach Alpha Centauri within 100 years using existing or near-future technologies. The four finalists presented their ideas in July 2015. The design submitted by the team from the University of California, San Diego, subsequently became Breakthrough Initiatives' Breakthrough Starshot project. The solar sail would be made of a single layer of the high-performance material graphene — a single layer of carbon atoms — and be 29.4 kilometres in diameter. It would require at least a 100-Gigawatt laser to accelerate a 2,750-kilogram spacecraft to 5 per cent of the speed of light, taking it to Alpha Centauri in about a hundred years. This laser could not be on the Earth as our atmosphere would be a hindrance, so it would be put at one of the gravitational balance points near the Earth. It would get its energy from solar panels and they would have to be enormous. Despite the optimism and publicity surrounding the Breakthrough Starshot project, this is not going to happen by 2069. The technical challenges are too formidable.

It is perhaps a crazy idea to get a microchip probe to the nearest star system and get information back but the prospects of such speedy travel within our solar system are tantalising. For example, it would take just a few days to reach Pluto, rather than the nine-and-a-half years that it took NASA's New Horizons mission. We could spray the microprobes into regions of the outer solar system and would be able to reach with ease any newly discovered objects coming into the solar system from interstellar space. In addition, the laser used to push the microchips could be used against a threatening asteroid as it would be able to ablate and deflect it.

Nearly 400 years ago astronomer Johannes Kepler observed comets had tails that always point away from the Sun, possibly pushed by sunlight, and suggested that vessels might likewise navigate through space using appropriately fashioned sails. Tennyson put it well:

> For I dipt into the future, far as human eye could
> see,
> Saw the Vision of the world, and all the wonder
> that would be,
> Saw the heavens fill with commerce, argosies of
> magic sails,
> Pilots of the purple twilight, dropping down with
> costly bales ...

THE MAN FROM SULAYMANIYAH

———————•———————

There is not a great deal of rocky debris left in our solar system compared to what there was. If you were able to gather together all the bodies of the asteroid belt situated between Mars and Jupiter it would only be about 5 per cent of the mass of the Moon. Yet their total surface area is about the same as that of the Moon, or Africa. This large surface area means that their metallic and mineral wealth is accessible. And some of them come quite close to us.

The first asteroid was discovered in 1801. It was named Ceres and soon many more followed. As of February 2020, the Minor Planet Center had data on almost 858,000 objects in the inner and outer solar system, of which about 542,000 had enough information to be given numbered designations. Ceres, at 1,000 kilometres across, is larger than many of the moons in our solar system. These objects are the wreckage of disintegrated small planets or debris that never formed any kind of larger body. They tell us much about our solar system's history.

Moreover, because of their vast mineral wealth they could play a critical economic role in the settlement of outer space.

The three broad composition classes of asteroids are C-, S-, and M-types. The C-type (chondrite) asteroids are the most common and probably consist of clay and silicate rocks. Dark in appearance, they are among the most ancient objects in the solar system. The S-types ('stony') are made up of silicate materials and nickel-iron and the M-types are metallic (nickel-iron). Their compositional differences are related to how far from the Sun they formed.

The first asteroid to be visited was the main belt asteroid Gaspra when the Galileo spacecraft flew by it on its way to Jupiter in October 1991. It was an irregularly shaped object roughly 19 kilometres long by 11. Some speculated that its origin was a dramatic collision between two larger bodies. In August 1993 Galileo passed the 50-kilometre-long Ida, which had a 1-kilometre-sized moon of its own, called Dactyl. This tiny moon's motion allowed the density of Ida to be estimated and it turned out to be about two-and-a-half times that of water. The first time we obtained a prolonged look and not just a flyby was when the Near Earth Asteroid Rendezvous (NEAR) mission reached and orbited Eros in January 1999. NEAR stayed at Eros for more than a year before landing on it. Subsequently the Japanese Hayabusa mission not only landed on the asteroid Itokawa but returned to Earth with samples in 2010. On the way to its famous encounter with comet 67P/Churyumov-Gerasimenko in 2014, the European Space Agency's Rosetta spacecraft passed asteroids Steins in 2008 and Lutetia in 2010. NASA's Dawn spacecraft orbited the large asteroid Vesta from 2011 to 2012, and then travelled to another, Ceres, in 2015.

The Japanese Hayabusa 2 and NASA's OSIRIS-REx missions were launched in December 2014 and September 2016 to explore the carbonaceous asteroids Ryugu and Bennu. Hayabusa 2 arrived at Ryugu in July 2018 and released the small European Mobile Asteroid Surface Scout (MASCOT) lander to collect samples. OSIRIS-REx reached Bennu in December 2018 and it will land on it in July 2020 and collect samples, which it will return to Earth in September 2023.

The vast mineral wealth, and potential profit, of these chunks of rock and metal has led many to think of mining them. It's been estimated that a kilometre-sized asteroid with a mass of a few billion tonnes would have 250 million tonnes of iron, 40 million tonnes of nickel, 3 million tonnes of cobalt and 8,000 tonnes of platinum, and lots of other metals in smaller quantities, let alone oxygen and carbon. This precious cargo is not buried deep in the ground as it would be on Earth but all within 500 metres of the surface. The proposition is therefore a straightforward one: how to get the resources out of an asteroid and bring them to the Earth–Moon system.

There are many asteroid mining companies today, almost all of them small speculative ventures with few resources, hoping to capture significant financial backing so that they can move on to do something more than proselytise. Most will fall by the wayside because it is too early for them and they won't be able to sustain themselves until better times come along. Today if you want to make money out of mining you had best stay on the Earth to do it. In the future, when sending probes to asteroids becomes easier, possibly due to the availability of liquid hydrogen and liquid oxygen from the Moon, the interest will pick up.

There could be far more money to be made if the metals were refined at the asteroid and then brought back, reducing transport costs. This would involve landing a spacecraft on a body with little gravity, anchoring it into place and drilling into the surface. Then the rock would have to be heated to extract the metal. On many asteroids this will not be possible since there is evidence that many of them are loosely bound rubble piles that would require little disturbance to turn them into clouds of debris, which may or may not be a good thing for potential prospectors. Processing ore in space will also be a problem and nobody has any firm ideas about how to do it at present. What we need is a long stay on the surface of a suitable asteroid and many experiments.

But someone has to start somewhere. In 2012 billionaire entrepreneurs announced a plan to mine asteroids and established a company called Planetary Resources. Its advisors included film director and explorer James Cameron and investors included Google's then chief executive Larry Page. Looking back on what they said just a few years ago is like reading science fiction.

Their plan was to create a fuel depot in space by 2020 by using water from asteroids, splitting it into liquid oxygen and liquid hydrogen for rocket fuel, just like the plans for the Moon. It could then be shipped to Earth orbit for refuelling commercial satellites or spacecraft. Although the timescales are obviously wrong this is not a bad commercial proposition for those with deep pockets and a long time to wait. The availability of water, either from the Moon or asteroids, has been estimated by some to be a potentially $200 billion business in the next 30 years. The large commercial satellites of the future will probably have refuelling capabilities so that when these profitable satellites run out of fuel they won't have to be shunted into a graveyard orbit but can be visited by a refuelling spacecraft and given a new lease of life.

This would be the start of a space economy that would allow for the eventual mining of asteroids. Another venture called Deep Space Industries started in 2013. It hoped to be prospecting asteroids in a few years and by 2016 returning samples to Earth, with actual mining taking place in 2023.

Exemplifying the mood at the time, in 2012, the NASA Institute for Advanced Concepts announced the Robotic Asteroid Prospector project. It suggested that the largest asteroid, Ceres, could become the main base and transport hub for future asteroid mining infrastructure, with its refined ores transported to Mars, the Moon, and Earth. It was an impressive study but that was all.

There are many candidate asteroids that could be mined. Ryugu and Bennu have been visited, as will the fascinating asteroid Psyche in 2026 by a mission to be launched in 2023. Psyche is one of the ten most massive asteroids and is over 200 kilometres in diameter. We have only poor, indistinct images of it taken by the world's largest telescopes. Some think it might be the remaining core of a protoplanet that was almost destroyed when the solar system was young. If so it would be a very interesting object to study as this core would be very enriched in heavy metals. It contains about 1 per cent of the mass of the asteroid belt.

Despite the premature hype, the mining of asteroids will come good in the long-term, even if it will not be at the scale predicted by some. Resources on Earth will eventually diminish for certain metals but it would only take the mining of a few asteroids to remedy that. For example, it's estimated that an asteroid like Psyche could provide more nickel-iron than we could ever conceivably need. Platinum is rare on Earth but the right asteroid just 30 metres long could contain tens of billions of pounds' worth of the stuff at today's prices. Space mining will take place, I have no doubt, but I fear many people will lose their money in the short-term. When it is done it will not be a widespread gold or heavy metal rush but targeted and limited, and for a few in the right place at the right time, highly profitable. If it flourishes it will need new markets and customers and they will not come from Earth, but that is for the centuries to come.

Asteroids might be bountiful, but they are certainly dangerous and the threat they pose to life on Earth has been well aired in recent decades by books and lurid science fiction films. A catastrophic impact event is possible, indeed probable, if mankind survives for a long enough period on Earth.

Every day some 8 million meteors enter the Earth's atmosphere. Most are small and burn up, but some become fine dust. High-flying aircraft can collect it in specially designed scoops. You can drag a magnet across the sea floor and pick up micrometeorites; you can detect them in ice laid down at the south pole thousands of years ago. But not all meteors burn up: some are large enough to survive the fiery entry and reach the Earth's surface. Hundreds of meteors strike the Earth each year, but fortunately large impacts are rare. Rare, that is, in a human lifetime, but inevitable over geological timescales. There is no record of anyone having been killed by a meteorite, as far as we know, though one story says that a dog was killed in 1911 at Nakhla in Egypt. It is estimated that the Earth receives a million kilograms of extra-terrestrial material each year. This may seem large but it's an insignificant contribution to the total mass of the Earth – about one billion billionth.

An object weighing several tonnes will strike the Earth every few hundred years. Larger impacts are rarer. In March 1989 an asteroid as large as an aircraft carrier crossed the Earth's orbit six hours after we had passed the same spot. It would have struck with the force of 2,000 one-megaton hydrogen bombs. Fortunately, it missed us but someday one won't. On 30 June 1908 a large body collided with the Earth in remote Siberia. This is thought to have been the nucleus of a small comet: just as nasty. The chances are small in the next 50 years but a certainty eventually. In 2018 the B612 Foundation that studies the subject said, 'It's 100 percent certain we'll be hit, but we're not 100 percent sure when.' A collision 66 million years ago between the Earth and an object about 10 kilometres wide is thought to have caused the Cretaceous–Paleogene extinction event, considered responsible for the extinction of most dinosaurs. Someone once quipped, 'The dinosaurs died out because they didn't have a space program.' Films and novels tell us what needs to be done: destroy the threatening asteroid or nudge it so that it will miss us. At the moment we are unprepared.

Most deflection strategies require years or decades to work and we haven't tested any of them. The sooner you get to a potentially hazardous asteroid the less of a nudge you have to give it. Asteroid 9942 Apophis will pass by the Earth in 2029 and has a one in 10,000 chance of passing through a region of space dubbed the keyhole. If it does it could set up an impact in 2035 or 2036. It's 370 metres wide, not an extinction-level threat, but potentially highly devastating. If we were able to change its velocity by just a millionth of a metre per second in the next few years, we would prevent it from entering the gravitational keyhole. Unfortunately, we are not set up to do this so we, all of us, will have to take our chances. I bet it will be OK this time.

We keep a careful watch on the skies. The Catalina Sky Survey project, operating from two telescopes in Arizona, looks for potentially hazardous objects, as does the Spacewatch initiative from another telescope in Arizona and there are a handful of other projects. They all feed into an Asteroid Terrestrial-impact Alert System which is funded by NASA and operates out of the University of Hawaii. The European Union supports the NEOShield project investigating ways to prevent a collision.

It's a gamble. Life on Earth has always been a gamble. Humans and their ancestors have gotten away with it for millions of years and it could be millions more years before we need to take action, and by then our planetary powers will be much greater. It's not impossible that such a threat will emerge in the next 50 years. Nonetheless, we should carry out tests and develop techniques to nudge an asteroid in its orbit and work out a way to blow it up if that doesn't work. The statistics say it's unlikely but in the back of my mind I can hear the words of the fictional detective Charlie Chan: 'Strange events permit themselves the luxury of occurring.'

There are claims that people have been hit and killed by meteorites but the historical records do not prove it. However, this may change because recently manuscripts were discovered recording that a meteorite hit and killed a man and left paralysed another on 22 August 1888 in Sulaymaniyah, in the Kurdish region of Iraq. The claim is based on three manuscripts written in Ottoman Turkish that were found in the General Directorate of State Archives of the Presidency of the Republic of Turkey. The event was reported to Abdul Hamid II (34th Sultan of the Ottoman Empire) by the governor of Sulaymaniyah. It seems that the meteorite fell on a hill near Çişane village. Some crops and fields were also damaged. These reports provide the most convincing evidence so far of the earliest case of a meteorite causing death and injury to humans.

We do not know the name of the unfortunate man from Sulaymaniyah but we do know he was not the first. Throughout our evolution there must be many more who were killed by impacts from space, just as many were killed by volcanoes and tsunamis. The threat from the leftovers of planetary formation is still the same as it was in 1888 but since then we have realised those bodies hold promise as well.

A DIMLY LIT WORLD

The ice giants Uranus and Neptune are neglected worlds, having only been visited once, and for over 40 years scientists have wanted to go back. They have an excellent scientific rationale for there is much to study and undoubtedly discover on these worlds, not least clues to the birth of the planets and how they may have moved considerably in their orbits in those early times. But somehow it has never worked; these planets have made it onto lists of desirable missions but never to the top of the list. It hasn't even been the case of waiting for one's turn, which is a strategy followed by other proposers of space missions. One scientist keen on exploring these worlds told me that the planets have passed a threshold and have become stuck. The sole visit to date was an expensive mission and a follow-up mission would be expensive as well, but these worlds are so far away and other missions judged more enticing have always taken precedence. NASA has recently shown signs of wanting to go back but the European Space Agency has said no. It does not foresee any missions to the ice giants before the 2040s. This is still within our 50-year purview, but a long way off for the scientists of today.

The power of the rockets we have available and the orbits of the planets limit the options for going back to these worlds. For Uranus the optimal launch window is 2030–34 and for Neptune it's even tighter: 2029–30. Both would require a velocity boost from a Jupiter flyby. In the best case it would take about twelve years to reach Uranus and a year or two longer to get to Neptune. As the money, rockets and mission slots are essentially set for the next decade it seems highly likely that this opportunity will be missed. It will be well over a decade before the next one. This puts the best case for a spacecraft arriving at the ice giants at the mid-2050s.

Uranus and Neptune may not have the allure of Jupiter and Saturn or the urgency of Mars and the Moon, but they are still fascinating, essential worlds that if we neglect to study from close quarters, we will never be able to say we understand our solar system. A sea change is needed in our regard for them.

Following its encounter with Jupiter and Saturn, Voyager 2 headed to Uranus, to the cold outer realm of the solar system, the place where Halley's comet turns around and heads back towards the distant Sun, the place where sunlight is hundreds of times weaker than it is on Earth.

We knew very little about Uranus before Voyager 2 arrived. On rare occasions it is barely visible to the naked eye. William Herschel discovered it on 13 March 1781. He saw a tiny bluish disc, 'a curious either nebulous star or perhaps a comet', but when it moved against the background stars he knew it must be a planet – the first one discovered since antiquity. Until 1985 none had seen it significantly better. With a diameter of 80,000 kilometres, only its atmosphere can be seen from Earth. The Hubble Space Telescope fares better, seeing bright clouds in its blueish atmosphere, just as Voyager did.

The rotation of Uranus is unique in the solar system, illustrating its violent past. Its spin axis is not at right angles to its orbit as is the case for all the other planets but aligned with it. This means that Uranus orbits the Sun 'on its side'. It's thought that a collision with a body the size of Earth could have knocked it over just as the planets were being formed. This results in some quite distinctive seasonal effects during its 84-year orbit. Each pole gets 42 years of almost continual sunlight or darkness, followed by the reverse. It has 27 known moons, all small – its moon system is the least massive among the giant planets – and a ring system composed of fine, dark particles.

Voyager 2's data is all we have from close up. It sent its images back using a transmitter with all the power of a dim lightbulb across 3 billion kilometres of space. Voyager 2 was heading towards a planet whose rotation period was not even known. In July 1985, half a year before close encounter, the first images were returned; already they exceeded anything that was possible from Earth. On 5 November the interplanetary cruise phase of the mission ended, and the observatory phase began. The photographs of Uranus got larger and showed ... nothing. Absolutely no detail, not even a cloud belt or a bright spot. This cast doubt on the observations made by ground-based astronomers which indicated an equatorial band of lighter coloured clouds. Perhaps, wondered some, the bands of Uranus will go the same way as the canals of Mars. Bland Uranus was surrounded by little white dots – its moons – and when the disc of the planet was overexposed the faint, dark rings discovered some years previously were revealed.

Because of its axial tilt, Voyager was to fly through the Uranian system rather like a dart towards a bullseye. Close encounter, at 80,000 kilometres above the cloud tops, would take just a few hours and Voyager was well prepared. Its onboard computer programs were rewritten, tested, simulated and the encounter rehearsed many times. This was vital. It would take the signals over two-and-a-half hours to reach Earth from Uranus and by the time the pictures started coming out of JPL's computer the close encounter would be almost over. Voyager was on its own.

The first new moon to be found was seen on 30 December, then a pair of moons was found shepherding the so-called epsilon ring, then a tenth ring was found. Details started to be visible on the disc: there was the banded pattern of clouds the ground-based astronomers had seen, appearing not unlike the low-contrast features on Saturn. The south pole, which Voyager was approaching, was darkish and surrounded by a band. Ten days before closest encounter the visible marking seen in the atmosphere enabled scientists to determine the planet's rotation period – seventeen hours fourteen minutes. Curiously, the wind speed appeared to be increasing with altitude, the opposite to what was expected. Aurorae were detected near the dark pole and something termed the electroglow was also seen: this faint glow of the atmosphere was caused by low-energy electrons from the magnetosphere striking the atmosphere. No one knew how.

A reason proposed for why Uranus was so bland suggested that because of its low temperature, methane ice crystals, which would have made the world more colourful, only form deep in the atmosphere, out of sight and not near the cloud tops as is the case for Jupiter and Saturn. Another clue to what's happening is the banding of the atmosphere. Significantly, the bands are aligned with the planet's rotation, not with solar heat, showing that it's rotation that dominates the flow of gases. The planet is almost all the same temperature, from its sunlit pole past the equator to the night-time pole.

Before Voyager's encounter scientists had no real evidence that Uranus had a magnetic field. There were a few tense days when Voyager's magnetometers were silent, well after the point when some scientists had predicted they would start registering a magnetic field. But just ten hours before closest approach Voyager passed the bow shock – the region where the solar wind ploughs into the Uranian magnetic field with great turbulence: Uranus did have a magnetic field, an astonishing one. Its magnetic axis was tilted and offset from the centre of the planet. Scientists thought that it was a freak but when Voyager got to Neptune it found that its magnetic field was almost the same.

To use Uranus's gravity to get on course for Neptune Voyager would have to pass very close to Miranda, the smallest of the major moons. Because of this closeness and the speed of Voyager, normal imaging techniques would have blurred the pictures. To compensate, engineers programmed Voyager to turn slightly to overcome the blurring; each image required its own compensation rate. Chaotic, jumbled and dramatic, Miranda was a marvel with regions resembling almost every solid body in the solar system. Scientists speculated that they may be seeing the exposed interior of the moon after it was shattered by impacts and reassembled itself, possibly many times. Miranda was an amalgamation of worlds from one end of the solar system to the other: it had the smooth contours of Mimas, the compression regions of Mercury and the grooves of Ganymede. Miranda also has Verona Rupes, the tallest cliff face in the solar system, 20 kilometres high. Given this tiny moon's tiny pull of gravity, it would take twelve minutes to fall from the top to the bottom. A daredevil feat for future sport enthusiasts.

On 10 February 1986 a two-and-a-half-hour firing of Voyager's thrusters gave it the extra few kilometres a second it needed to reach Neptune, 1.6 billion kilometres further out and three-and-a-half years in the future, its final planetary encounter. 'This is the last picture show, the last foreign shores we visit in the solar system,' said one scientist. Despite the anticipation of the Neptune encounter there was an air of sadness among the scientists at the Jet Propulsion Laboratory; after twelve years and 115,000 images, this was the finale. Voyager had travelled 7.1 billion kilometres to reach the fourth-largest planet in the solar system. They had measured births, deaths and marriages by it, and soon it would be over.

The formal encounter phase began on 5 June 1989 when Voyager was 117 million kilometres from Neptune. Already a great dark spot had been visible on Neptune for many months. More atmospheric activity was expected on Neptune than on

Uranus. Neptune has an internal heat source, a relic of its creation, whereas if Uranus has an internal heat source then it is very feeble. This means that despite being almost twice as far away from the Sun as Uranus, Neptune is almost the same temperature, and that heat should drive weather systems.

By 3 August new moons, small, battered objects, had been discovered and the previously known arcs of rocky debris around the planet were beginning to be visible. About one week before the flyby Voyager detected powerful bursts of radio energy from charged particles in the planet's vicinity. It came as a surprise that Neptune's magnetic field is considerably weaker than that of Uranus.

A few days before close encounter, Voyager 2 celebrated its twelfth birthday in space. Its final ever course adjustment was made in a manoeuvre called TCM-20 that changed its velocity slightly, altering its aim point 146 kilometres further from Neptune and 706 kilometres nearer its moon Triton. Scientists estimated that Voyager would arrive above Neptune's cloud tops four-and-a-half minutes early. Soon the incomplete arc around the planet had become a ring.

The outer layer of atmosphere of both Uranus and Neptune consists of hydrogen and helium with a little methane and a trace of ammonia. Going deep into the atmosphere, the gaseous layer gives way to a denser layer composed of water, methane and ammonia. About two-thirds of the mass of Uranus and Neptune may be in this intermediate layer. At the core, at a pressure 20 million times that of Earth's atmosphere and a temperature of 7,000°C, there is a rocky core. In a sense there's an Earth at the centre of this planet.

But it was Triton that was undoubtedly the star of the Neptune encounter, a fitting last shore of the solar system, as one scientist put it. Triton is the only major body in the solar system that orbits its parent body in the direction opposite to the spin of its parent body. It's thought that it was once free and was captured either by being hit by a smaller moon of Neptune or by drag in the extended atmosphere of the proto-Neptune billions of years ago. For Voyager's engineers, Triton was a bit of a surprise: it soon became clear that it was smaller and brighter than had been expected. A last-minute change reduced the exposure of Voyager's cameras at Triton by 50 per cent, otherwise some of the pictures would have been overexposed.

Triton has an atmosphere composed of very cold methane gas, which had previously been detected by Earth-based telescopes. It has some similarity to the atmosphere of Mars. As on Mars, the sublimation of the sunlit polar cap is important as a source of atmospheric gas and a source of pressure differences to propel winds across the barren landscape. As Voyager's radio beacon passed behind Triton, scientists concluded that the atmospheric pressure was 100,000 times less than that of Earth. Most of Triton's northern hemisphere remained in shadow during the flyby. Large tracts of the equatorial regions were mottled; fault lines displayed a regular rectangular pattern. Triton had a polar cap of frozen nitrogen and methane. Puzzling dark streaks were seen near the south pole, similar to streaks on Mars caused by wind-blown deposits – however, Triton's atmosphere is too thin to move fine dust.

Frozen lakes, edged with terraces, seem to indicate that water slurry, forced up by volcanic heat, has erupted from the interior and spread over the surface, freezing before it could retreat. Similar formations are seen carved out in solidified magma in Hawaiian volcanoes. A region described by geologists as 'cantaloupe terrain' showed repeated local melting and collapse of the crust. Cracks some 35 kilometres across were seen filled with fresh ice welled up from the interior. There were strange raised features with dark centres and white collars.

But perhaps even more remarkable than the obvious evidence of past activity on the moon is that Triton is, in a small way, an active world. Like Io, Triton has volcanic activity in the form of geysers. Two huge plumes of nitrogen were detected near the south pole on the Sun-facing side, rising 8 kilometres above Triton and blown 150 kilometres downwind. Some believe that subsurface liquid nitrogen is heated and is released explosively through vents. Perhaps this is caused by the greenhouse effect trapping heat beneath a surface layer of nitrogen ice.

Triton is a dimly lit world. Dawn on Triton takes about 25 seconds but as the Sun begins its three-day passage across the sky it brings little warmth. Seasons on Triton are peculiar to put it mildly. It keeps the same face towards Neptune during its 141-hour orbit, Triton's orbit inclined to Neptune's equator by 21°. This means that as Neptune carries Triton around the Sun the tilts of the planet and its moon can either add up or cancel. If they add then sunlight shines on one of Triton's poles for 50 years; when they cancel it shines on

Triton's equator for 100 years. This peculiar cyclicity of seasons built up to a relatively warm summer in 2007. But in a hundred years Triton will be having minimum summers and minimum winters, the equator will bask in sunlight and both poles will be cold. Its surface may have changed as nitrogen ice is redistributed; it may look quite different.

For a return to this fascinating world, NASA is considering a mission called Trident – a single flyby mission to explore Triton. Triton has active resurfacing – generating the second youngest surface in the solar system – with the potential for erupting plumes and an atmosphere. This is coupled with an ionosphere that can create organic snow and the potential for an interior ocean. Trident is only partly what ice giant advocates want – they would have preferred an orbiter – but they will welcome Trident, if it beats off its three other competitors for funding.

At the edge of our solar system orbits the tiny ice world of Pluto and its moon Charon, only visited once when in July 2015 the New Horizons probe flew past the system. Since then scientists have wanted to go back. The Principal Investigator of New Horizons, Alan Stern, has suggested an orbiter like the Cassini craft that orbited Saturn. It could launch about 2030 (the 100th anniversary of Pluto's discovery) and use Charon's gravity to adjust its orbit as needed. It's a good idea as Pluto is an intriguing world but the time is not right politically and I would think advocates for a return will have to resign themselves that it will be more than a decade, probably longer, before we go back, given all the other exciting missions in prospect.

POSADKA

———————•———————

I have only seen Mercury a handful of times. Most people haven't seen it at all. Most astronomers probably haven't. Because it's the closest planet to the Sun, 58 million kilometres from it, it never strays far from the Sun in the sky. Like Venus, it appears as an evening or a morning object. To see it you have to look low down on the horizon and know where to look. Little detail can be seen on it, even through the world's largest telescopes.

Only two spacecraft have visited Mercury. Mariner 10 first flew to Venus and used the planet's gravity to reach tiny Mercury. It was the first practical demonstration of the gravitational slingshot effect whereby a spacecraft can gain speed by flying past a planet. After its encounter with Venus Mariner 10 entered solar orbit and encountered Mercury on three occasions, making passes of the planet in March and September 1974 and March 1975, observing the same area of the planet each time. In all, 75 per cent of the planet was mapped. MESSENGER arrived in 2011 and studied the planet until 2015. MESSENGER found large amounts of water in Mercury's thin extended atmosphere. It also saw much evidence for ancient volcanic activity. In its mineralogical mapping it found carbon-containing organic compounds and water ice inside permanently shadowed craters near the north pole, just like the Moon.

Most astronomers expected Mercury to look very much like the Moon and they weren't disappointed. At first sight it was very Moonlike: mountains and craters and vast plains where geologists believe volcanic lavas once flowed. A closer look, however, reveals some distinctly Mercurian landforms. There are hundreds of long cliffs stretching across the surface, which geologists call lobate scarps. These cliffs, some more than two kilometres high, 24–480 kilometres in length, are believed to have formed when the crust of Mercury shrank and wrinkled as its interior heat source waned soon after its formation.

The largest structure is the Caloris basin, some 1,300 kilometres in diameter, formed by a gigantic impact several billions of years ago. Unlike basins on the Moon, its floor displays closely spaced ridges showing both radial and concentric fractures due to shock waves from the impact. On the opposite side of the globe to Caloris is a peculiar terrain of hills and valleys. Scientists believe that the shock waves from the impact travelled both ways round the planet to come to a focus at the antipodean point, creating new landforms. It must have been quite an event.

The origin of the plains is a major mystery. They are probably volcanic in origin and were likely formed near the end of the heavy meteorite bombardment of 3.8 billion years ago. If they were formed at this time then they are older than the lunar maria. The energy from these meteorite impacts could have melted the surface of the planet, and a large impact could have melted the entire planet. Geologists think Mercury experienced more extensive volcanism than the Moon after such impacts.

The Sun looms over the planet, raising the temperature to 480°C at high noon. Lead and zinc would flow like water. With no atmosphere to conduct the heat to the nightside of the planet the temperature there falls to minus 180°C just before dawn. Mercury displays the greatest temperature range of any planet. Despite this wide temperature variation it seems that the surface is a good insulator, for only 1 metre below the surface the temperature is an even 77°C whatever the time of day.

One of the greatest surprises of the Mariner 10 mission was Mercury's magnetic field. Although only about 1 per cent as strong as the Earth's its interaction with the solar wind does allow Mercury to form a magnetic sheath around the planet, a magnetosphere. A compass needle on Mercury would point north. But why is it there? Scientists had thought that Mercury's slow rotation would inhibit the formation of a magnetic field because such fields are generated by fluid motions in the planet's core. If the planet doesn't spin quickly then it's an indication that motions in the core won't be strong either. Moreover, the lobate scarps indicate that Mercury's internal heat source faded long ago and so it was thought that the core must be frozen solid and unable to generate a magnetic field through its motions. But the weak magnetic field which has been detected must mean that some part of Mercury's core is still molten. Scientists speculate that the presence of some impurity such as sulphur in the core could enable it to be partially molten at unusually low temperatures.

Mercury has a heart of iron. Over 70 per cent of the planet's mass may reside in its metallic iron core, which occupies about 75–80 per cent of the entire planet. Mercury's rocky coating is but a thin layer around an immense iron core. This makes it quite different from the other terrestrial planets, which have modest iron cores. It raises an interesting question about Mercury's origin. Did Mercury form in the hottest region of the solar nebulae where it retained only the heaviest elements? Or has its exterior, where the lighter elements would reside, been removed due to some gigantic impact?

Despite its small size, Mercury has a very thin atmosphere. In 1978 the McMath solar observatory at Kitt Peak detected tenuous sodium emission. Concentrated at high latitudes, it varied from day to day. It is possible that Mercury's magnetic field may funnel charged particles from the solar wind down to the surface where they dislodge sodium atoms. A few years later evidence of potassium was also detected. Mercury's atmosphere is at a pressure one-trillionth that of the Earth's. It's very nearly a perfect vacuum – a vacuum so good that the atoms in its atmosphere rarely collide.

Mercury may have formed when about two dozen bodies coagulated to form a proto-Mercury about twice the mass of Mercury today. It may have then collided with another body and the proto-Mercury melted, causing the heavier elements like iron to sink towards its core. One theory even suggests that this proto-Mercury was orbiting the Sun between the orbits of Earth and Mars and a collision stripped it of its outer shell, leaving the iron-rich core behind. Gravitational encounters with Mars and the Earth, and perhaps other bodies orbiting the Sun during those early times, displaced Mercury to its present orbit. Alternatively, the heat given off by radioactive elements could have melted it soon after its formation, leaving the iron to sink to its core. Whatever its origin, it's clear from the lobate scarps that Mercury expanded, fracturing the thin crust and allowing lava to ooze out to form the intercrater plains. As it cooled the planet shrank a few kilometres and the lobate scarps formed.

There is one mission planned for Mercury. BepiColombo is a joint mission of the European Space Agency (ESA) and the Japan Aerospace Exploration Agency (JAXA). It consists of two satellites launched together: the Mercury Planetary Orbiter (MPO) and the Mercury Magnetospheric Orbiter (MMO). It was launched on an Ariane 5 rocket in 2018 with an arrival at Mercury planned for December 2025, after a flyby of Earth, two flybys of Venus, and six of Mercury.

The Russians have proposed Mercury-P, a lander. The P is for *posadka*, which means landing. It's going to be the 2030s before it's launched. It has also been suggested that Mercury would be a good place to build and launch solar sail spacecraft, because sunlight at Mercury is so intense.

With the human challenges of the Moon and Mars before us, I cannot see that humans will go to Mercury, even though it would involve similar challenges to going to the Moon as there are few volatile elements, no atmosphere and the surface gravity is lower than Earth's. The problem is getting there, and the even bigger problem is that being so close to the Sun it receives almost seven times the solar flux received by the Earth–Moon system.

No doubt there will be many future probes that will visit Mercury, some attempting landings, but I fear its fate is to be a lonely outpost, forever seared by the harshness of the Sun.

ADJACENT AND
IDENTICAL

———•———

They say Venus once had warm oceans that for millions of years lapped the shores of its continents – with no moon in its skies the tides on Venus would have been slight. But two-and-a-half billion years ago all that changed as the planet became hotter and hotter and there came a time when the last drop of water evaporated from the oceans. Venus has changed into a baking hell because of the greenhouse effect, and while all this was happening a bright object periodically appeared in its night sky, only 42 million kilometres distant – a similar planet, later to be called Earth, that escaped this fate.

Venus is now very different from what it once was. Being closer to the Sun than the Earth, it never strays far from it in the sky and is often seen as a brilliant evening or morning object. Only the Sun and Moon are brighter. Venus's brilliance comes from its perpetual cloud cover that reflects a lot of sunlight and hides what is beneath. Once astronomers had romantic notions, based on very few facts, that underneath there might be lush jungles teeming with all sorts of life, or perhaps a planet-wide ocean of hydrocarbons peppered with occasional islands. But that changed in 1956 when a radio telescope was turned towards the planet. It was able to detect radiation not stopped by the clouds and it showed that the surface was very hot, nearly 500°C, hotter than the hottest household oven. There were no oceans, indeed no moisture at all, just baked rocky plains and mountains. The radio observations were also able to solve another mystery, the rotation period. This was found to be 243 Earth days, longer than it took Venus to orbit the Sun. As a result, as seen from Venus the Sun rises in the west and sets in the east, taking 118 days to do so. The Earth and Venus are in what is termed orbital resonance, Venus presenting the same face towards the Earth each time it is closest to our planet.

Many missions have visited Venus. The first successful one was Mariner 2, which passed Venus in December 1962. It confirmed, as did later missions, that the temperature beneath the clouds is about 470°C and the pressure at the surface is about 90 times greater than that on Earth – that's the pressure 1 kilometre beneath the surface of the Earth's oceans, so on the Venusian surface you would be suffocated, baked and crushed. Nonetheless, spacecraft launched by the Soviet Union have survived these conditions, briefly, and have landed on the surface to send back a few close-up pictures of the rocky terrain. Other spacecraft, including NASA's Pioneer Venus Orbiter and the Soviet Union's Venera 15 and 16, have used radar from orbit to make low-resolution maps of what lies beneath the clouds.

Near the top of the atmosphere the wind speed is some 300 kilometres per hour and winds whip around the planet in a few days. The atmosphere itself is 96 per cent carbon dioxide with some nitrogen, water vapour, argon and carbon monoxide, as well as some small quantities of hydrochloric and hydrofluoric acid. Sixty kilometres above the surface there are clouds of sulphuric acid droplets which are circulated up and down; the droplets break up as they fall and reform high in the atmosphere so not a drop of this corrosive mixture ever reaches the surface rocks. Deeper down, the atmosphere is hot and dense but clear, with about as much sunlight as an overcast day on Earth.

Pioneer Venus Orbiter arrived at Venus in 1978 and used its radar to map about 90 per cent of the planet. It found three types of terrain. Most of the planet is covered in rolling plains with some lowlands and a smaller fraction of highlands. Beneath its clouds Venus has continents, the largest one being Aphrodite Terra, the size of Africa. Smaller but more complex is Ishtar Terra, which seems to contain the deepest valleys, about 4 kilometres deep. Ishtar Terra is about the size of Australia and has one of the solar system's highest mountains, Maxwell Montes, at 11 kilometres high some 2 kilometres higher than Everest. Its summit, called Cleopatra Patera, is marked by a large double depression, probably of volcanic origin. On the southern flank of Ishtar Terra are giant cliffs which descend to vast lowland plains. To the east of Maxwell Montes is a complex region of ridges and groves.

The Earth's surface is divided into plates that move in relation to each other. Some plates duck under others in a process called subduction. This process, the cause of continental drift, is the Earth's way of losing heat from its hot interior. Pioneer found no equivalent globe-circling system of ridges and faults to that which on Earth marks the boundaries between plates.

The next probe, Magellan, was put together using a lot of leftover parts from other spacecraft. Its antenna was a Voyager project spare. Other pieces were cannibalised from the Viking, Galileo and Ulysses missions. Its most important instrument was what is called a synthetic aperture radar, which basically is a radar dish of a certain size that takes advantage of its movement in orbit around Venus to perform as if it were bigger. There was a similar radar package on Pioneer, but it couldn't see fine detail and was only able to map out the broad outlines of the Venusian surface.

Meanwhile the Soviet Veneras 9 and 10 survived the descent through the atmosphere and landed on the shores of an upland region called Beta Regio. They took pictures and analysed the rocks, showing them to be similar to basalt, like the rocks on the Earth's ocean floor and those that ooze out of Earth's volcanoes.

Magellan was released from the space shuttle on 4 May 1989 to begin the passage to Venus, which it reached in August 1990. It settled into an orbit that took it around the planet every three hours, nine minutes. When close to Venus, the spacecraft would point its imaging radar at the planet's surface to collect data. When farther away it would transmit its data to Earth. Each orbit produced a strip 20–25 kilometres wide and 15,000 long – scientists called them noodles.

The first images were of Beta Regio, which has ridges and valley floors where volcanic flows have hardened into a network of fractures and small craters.

The lack of impact craters shows that Venus's surface is young in geological terms, but no one had any idea how young until relatively recently. In data collected in 2010 by the European Space Agency's Venus Express spacecraft, which orbited the planet between 2006 and 2014, a series of hot spots were seen. They were identified as lava flows considered to be very young – that is, 250,000 years old. Two years later Venus Express detected spikes of sulphur dioxide in the atmosphere, suggestive of an active volcano erupting. Similar spikes had been detected in 1978, 1986 and 2006. A few years later an enterprising scientist looked at what happens to the volcanic rock olivine when it is subjected to the conditions found during an eruption on Venus. It was found to turn to iron oxide very quickly. Looking back at the lava flows thought to be 'only' 250,000 years old, traces of olivine could be seen. That it had not been changed to iron oxide suggested it was even younger. In fact, just years or tens of years. There are active volcanoes on Venus right now.

There is no shortage of proposals for a return to Venus. VERITAS (Venus Emissivity, Radio Science, InSAR, Topography, and Spectroscopy) is a proposed mission concept by NASA's Jet Propulsion Laboratory to map the surface at high resolution. Detailed knowledge of the topography would provide insights into Venus's tectonic and impact history. EnVision is a proposed orbital mission that would also use radar to map it. The concept was selected in May 2018 as a finalist to become the fifth medium-class mission of the Cosmic Vision program by the European Space Agency. The winner will be selected in 2021 and will launch in 2032. DAVINCI (Deep Atmosphere Venus Investigation of Noble gases, Chemistry, and Imaging) is a proposal for an atmospheric probe to Venus. It lost out in the 2015 round of proposals for NASA's Discovery Program but was proposed again in 2019 and shortlisted in February 2020. Before it reaches the surface, the DAVINCI probe would take the first ever photos of the planet's intriguing, ridged terrain. The most ambitious proposal is to get something to the surface and keep it working. Known as the Long-Lived In-situ Solar System Explorer, or LLISSE, it will have to withstand high temperature, high pressure, and a reactive atmosphere which contains a tiny amount of sulphur, which forms crystals on electronics. As it is, the designers can't see a way to put a camera on board.

Venus is for atmospheric scientists and geologists, a reminder of how the Earth could have developed if conditions had been just that little bit different. As it is, the Earth is a blue heaven compared to the scorching, oppressive heat of our nearest neighbour in space. The Greek writer Nikos Kazantzakis summed it up by writing that the doors of heaven and hell are adjacent and identical.

VISITORS FROM DISTANT STARS

———•———

On the fringes of the vast cloud of gas and dust that was to become our solar system, gases froze to become grains. Somehow, by processes we do not yet understand, these particles coalesced to form clumps of dust and ice in a huge cloud of perhaps a thousand billion comets some 50,000 to 100,000 times further from the Sun than the Earth. Comets are insubstantial things – all of them together would barely weigh as much as two Earths – yet they live in the vastness of the cold and dark in the outer solar system. Once in a while a nascent comet is drawn from the crowd, stirred by the gravitational pull of the planets or perhaps a passing star, and falls sunward. As it warms, its ices turn to gas and a dust-and-gas tail billows from it, forced away by the Sun's light and the solar wind. Most comets pass the Sun just once and return to deep space. Others get their orbits changed by gravitational encounters with the planets. These are the short-period comets, of which Halley's comet is one, returning every 76 years or so. When a comet is confined to the inner solar system, each passage past the Sun loses it dust and ice until it fizzles away, becoming a meteor stream. Observe a comet before this happens and you can read secrets written before the planets were formed.

I remember Halley's comet. For a while in 1985 many awaited it with great expectation. While it wasn't an impressive sight in the night sky it was a beautiful sight in binoculars and telescopes. But this return of the comet was the first to be visited by spacecraft. In all, six space probes converged on it. Two were Japanese, two Russian, one was a joint USA–European craft and the most daring of all was the all-European Giotto, named after Giotto di Bondone, the Renaissance painter who once depicted the comet as the star of Bethlehem in his painting the Adoration of the Magi.

Approaching the inner dust shroud of Halley at twice the speed of a bullet, Giotto's instruments were protected by two shields. Giotto's close pictures showed, much to astronomers' surprise, that most of Halley's nucleus is not active. Only a small fraction, 10 per cent, was covered by active gas-and-dust jets. It's estimated that every second Halley lost 20 tonnes of gas and 10 tonnes of dust, mostly from the jets. This is a small fraction of the total amount of gas and dust in Halley's nucleus. With a volume of some 500 cubic kilometres it contains at least 100 billion tonnes so there's a lot left for future returns. One surprise was how dark the material was – among the blackest stuff in the solar system. After Giotto encountered Halley it was left in solar orbit. But in July 1992 it flew by comet 26P/Grigg-Skjellerup. It's still orbiting the Sun, but its comet-hunting days are over.

The most dramatic mission to a comet was undoubtedly the European Space Agency's Rosetta, which along with its lander Philae moved alongside comet 67P/Churyumov-Gerasimenko in 2014–15, studying its topology and composition. In November of that year Philae landed on the surface, seemingly bounced, and was lost in a crevasse. It sent back a picture, while crippled and lying on its side, of nearby ice mixed with rock. In September 2016 the mission ended as Rosetta itself attempted a soft landing on the comet.

We have learned a lot about comets and can imagine what it would be like to stand on one. Most of the time life on a comet would be boring. When far from the Sun, little happens to disturb the tranquillity of its frozen landscape and clear view of the stars. Its pull of gravity is so puny that if you stood on a solid patch of rock you could leap 30 kilometres up into space and take over a week to come down again, seeing this frozen iceberg rotate a few times beneath you as you descended. Such a small body would be a place of strange effects. You could walk to the equator and throw snowballs against the comet's rotation and watch them float almost motionless in front of you; you could throw them away forever. It is only when the comet approaches the Sun that things start to happen. The ices start to turn into gas, mountains start to move, geysers erupt and the sky fills with light. From the comet flow tails of gas and dust perhaps millions of kilometres long. There will be many missions to comets in the next 50 years and perhaps the first to a comet or asteroid not born around our Sun.

'Oumuamua (Hawaiian for 'first arriving distant traveller') is the first known interstellar object to have passed through our solar system. It was discovered from an observatory in Hawaii in October 2017, 40 days after it had passed its closest point to the Sun on 9 September. When found it was about 33 million kilometres from Earth – 85 times as far away as the Moon – heading away from the Sun. It was a small object, approximately 100 metres wide and 1,000 metres long, making it a most elongated object. This led to speculation that it could be a derelict spaceship! It had a dark red colour, similar to objects in the outer solar system. It could be a remnant of a disintegrated rogue comet or a shard of an Earth-sized planet that broke apart when orbiting a distant star. It was tumbling, rather than smoothly rotating, and moving so fast relative to the Sun that there is no chance it originated in the solar system. It is now heading back out into interstellar space. Some astronomers have wondered if it could be caught up, but although there are various options to send a probe after it they will not be taken up.

In August 2019 another interstellar interloper was seen heading towards us from the direction of the constellation Cassiopeia. This time it really looked like a comet, coming in swiftly on a trajectory to and from the stars. It was called 2I/Borisov, the designation 2I standing for second interstellar object. In many ways it is similar to traditional long-period comets. It rounded the Sun on 8 December and made its closest approach to Earth soon after. It's leaving in the direction of the constellation Telescopium and calculations indicate it will not reach another star for a very long time.

It would have been much harder to send a spacecraft to reach 2I/Borisov than 1I/'Oumuamua but these two visitors have prompted some to suggest we should be ready and have an intercept spacecraft on standby in space. Calculations performed by a team at the Initiative for Interstellar Studies suggest that a two-tonne spacecraft could theoretically have been sent in July 2018 to intercept 2I/Borisov if launched by a SpaceX Falcon Heavy rocket. As it was, it was discovered too late for anything to reach it. NASA has said it may need at least five years of preparation to launch such an intercept mission. The Comet Interceptor spacecraft, supported by ESA and JAXA, is planned to launch in 2028 and will be positioned to wait at the Sun–Earth Lagrange gravitational balance point to wait for a suitable target. If nothing suitable is found after three years it will be sent to the best available cometary target.

Prompted by these two interstellar visitors, astronomers began searching the solar system to see if they could identify any interstellar object that might have become trapped here. After running computer simulations of how the solar system may have looked in its earliest days they found a coterie of them. They came from the Sun's siblings.

The Sun was not born alone but in a stellar nursery alongside hundreds, even thousands of others each a collapsing cloud of gas and dust with a young star at its heart and surrounded by rocky and icy debris in the form of comets. When close together this debris would have intermingled and comets been exchanged between the infant stars. Today several such bodies have been identified by their peculiar orbits. There are now nineteen candidates orbiting the Sun as part of a larger group called Centaurs which move between the giant planets. The strange trajectories of the Centaur objects have always puzzled astronomers. Not all of the candidates would have come from the same star and a mission to send a probe to examine and possibly land on an object not born of our solar system will prove irresistible over the next few years or decades.

The Sun's siblings are now scattered all across the galaxy and, who knows, many of them may have comets from our own Sun orbiting in their outskirts.

2069

———•———

We have choices to make about where we want to be in 2069. We could get there having made little progress in space and be content as long as the Earth reaps the benefits of its space programs with better telecommunications, better weather forecasts, better satellite navigation systems and other things that rely on space that will increasingly integrate themselves into our daily lives. But I contend that this is not an option for us. We have already passed a tipping point in the assimilation of space into our technological, military and political lives, and it has become a key element of national power. It is not like it was in the 1960s with the race between the USA and the USSR. This is longer term and with higher stakes. China is developing a national space strategy in preparation for the time, perhaps twenty years away, when it becomes the world's dominant economy. It is inconceivable that the US will not respond, or else it will place its national power at risk. It will expand its role in space and that is a good thing because it will mean in all of the major economies of the world there will be a greater investment in science and technology, and on the education of young scientists.

We could get to 2069 having settled for the level just above that we achieved with the International Space Station. We will have sent humans back to the Moon and have a thriving space tourism industry, especially in low Earth orbit. But if we confine space activity to this and operating satellites to low Earth and geostationary orbit, then space will be a reduced player in the global economy compared to what it could be. Alternatively, we could have thousands of people working in space. We could expand our domain as world population and GDP increases that will lead to increased competition for natural resources, on Earth and perhaps then on the Moon.

By 2069 it is possible that global political power will become more widely spread as the difference between the top nations narrows. Any conflicts between them will involve many aspects of society, the economy and the military and they will reach out into cislunar space.

You can't see what space will become 100 years after the first Moon landing just by looking at the missions to the planets and by how far humans will travel. As I said much earlier, we will take all we have with us, our hopes and fears will be the same ones we have always had, just transplanted into the solar system.

Some time between now and 2069 there will be a significant event in the history of our species. It will not be landing on the Moon again, or on Mars, or anything like that. It will be a more basic, more fundamental event, one that we all understand, one that will change everything.

The question of giving birth in space has long been discussed and is a staple of science fiction and tabloid newspapers. There are some that want to send a pregnant astronaut into space on a brief tourist trip and allow them to give birth before coming back – highly irresponsible in my view as we have seen what weightlessness can do to you and we can't be sure things would go well. I suppose two space tourists on a Virgin Galactic flight could start the process. But for me that would be bad taste and symbolic in the wrong way.

Will conception and birth take place on a space station, on the Moon, or during interplanetary transit? I've no doubt that in the centuries to come it will be all of these things but a birth on Mars would be symbolic of how we are changing as a species. Babies encapsulate all of our best hopes and dreams for the future, a slow relinquishing of the Earthly affairs of our personal times. Whatever we do in space, on the Moon and on Mars, we will pass it on to those to come. Given the possibilities outlined in this book I even think that the parents of the first space child are already among us. Every child should inherit the world. What about a child who inherits a brand new one? We will be a terrestrial species no longer.

Of course, we do not have to go to the Moon and Mars. We could spend our money and resources on Earth, facing all the problems we have. I do not believe that expanding into space means we must forego what we need to do on Earth. Missions beyond our planet only make sense in the wider spectrum of human endeavour. Only when we are mastering our bodies with new understanding and power to alleviate the diseases that scourge us, only if we tackle poverty, does a space program make sense. It's about something we can't do without if we are to thrive. Confidence. Lack of it strikes at our very heart and soul and the meaning of our lives. Without it civilisations wither from within.

By 2069 we will be sharing our space exploration with artificial intelligences – robots that can roll, crawl, swim and fly. They can be one big robot or many small ones, and they can explore where humans cannot. They look like drones, spiders, beachballs, wheels, and they are all programmed to move, sense and move again. They will be let loose on the Moon, scattered all over Mars, dropped into the smoggy atmosphere of Titan, stuck to asteroids, thrown into comets. When robots make robots our exploration horizon will expand rapidly. Perhaps it will not be us that meet ET but our AI emissaries.

Perhaps the timescales will slip? Will the next footprint on the Moon be 2024 or a few years later? What effect will the coronavirus pandemic have? If President Trump loses the 2020 presidential election, will President Biden continue the space initiative? Whoever is the president of the United States in 2021 the Treasury will have run up a colossal debt supporting industry during the pandemic and Congress may not be in the mood to support the 2024 deadline. I think the program will stay but slip a year or two perhaps.

One hundred years after Apollo 11 humans will have visited all the worlds we can. The gas worlds beyond Mars and the icy worlds beyond them will not beckon us as strongly as the new worlds we will find around other stars. Giant telescopes, on the Earth and in space, will send us more and more details of them. Decoding the light from these worlds may tell us if they have life. It is to these worlds that we will turn our attention in the next 50 years as we work out how to obtain an image of the evening sky of a world beyond our solar system.

At this moment there are active volcanoes on Venus; briny water flowing on Mars; methane rain falling on Titan; hot vents at the bottom of a dark ocean inside Europa; ice forming at the Martian poles; geysers on Enceladus and Triton; skylights on the Moon; astronauts on the space station, rovers on Mars; volcanoes on Io; someone training to walk on the Moon; and the first person on Mars is at college.

In 2069 the Earth will be in the evening sky of Mars, with the Moon visible through a telescope. The future Martians are making their plans. Soon they will be roving over its red dust, sometimes listening to the obligatory soundtrack that will accompany all its explorers and all its new horizons and peoples. 'Look at those cavemen go.'

BIBLIOGRAPHY

Howell, Elizabeth, and Nicholas Booth, *The Search for Life on Mars*. Arcade, 2020.

Zubrin, Robert, *The Case for Space*. Prometheus Books, 2020.

Pyle, Rod, *Space 2.0*. BenBella Books, 2019.

Davenport, Christian, *The Space Barons*. Public Affairs Reprint Edition, 2019.

Fernholz, Tim, *Rocket Billionaires*. Houghton Mifflin Harcourt, 2018.

Lewis, John S., *Asteroid Mining 101*. Deep Space Industries, 2014.

David, Leonard, *Moon Rush*. National Geographic, 2019.

Heiken, Grant H., David T. Vaniman and Bevan M. French (eds), *Lunar Sourcebook*. Cambridge University Press, 1991. (https://www.lpi.usra.edu/publications/books/lunar_sourcebook/pdf/LunarSourceBook.pdf)

INDEX

INDEX

ABOUT THE AUTHOR

Dr David Whitehouse is a former BBC Science Correspondent and Editor. He studied astrophysics at the world-famous Jodrell Bank radio observatory. He is the author of several books, including most recently the acclaimed *Apollo 11: The Inside Story*. He has written for many newspapers and magazines and regularly appears on TV and radio programmes. Asteroid 4036 Whitehouse is named after him.

www.davidwhitehouse.com